Lóegaire Humphrey (Ed.)

Harlan County Reservoir

Lóegaire Humphrey (Ed.)

Harlan County Reservoir

Nebraska, Republican River, Reservoir, Phillips County, Kansas

Claud Press

Imprint

Publisher:
Claud Press is a trademark of
International Book Market Service Ltd., 17 Rue Meldrum, Beau Bassin, 1713-01 Mauritius
Email: info@bookmarketservice.com
Website: www.bookmarketservice.com

Published in 2011

Printed in: U.S.A., U.K., Germany. This book was not produced in Mauritius.

ISBN: 978-613-7-41668-6

Contents

Articles

Harlan County Reservoir 1

Nebraska 4

Republican River 20

Reservoir 23

Phillips County, Kansas 33

Colorado 39

Republican City, Nebraska 72

Alma, Nebraska 74

Brownville, Nebraska 77

References

Article Sources and Contributors 80

Image Sources, Licenses and Contributors 82

Harlan County Reservoir

Harlan County Reservoir	
Location	Harlan County, Nebraska / Phillips County, Kansas, USA
Coordinates	40°04′15″N 099°12′41″W
Lake type	reservoir
Primary inflows	Republican River
Primary outflows	Republican River
Basin countries	United States
Max. length	7 mi (11 km)
Surface area	13250 acres (5360 ha)
Surface elevation	603 m (1978 ft)

The **Harlan County Reservoir** includes a dam and a reservoir of 13250 acres (54 km^2) located in Harlan County in south-central Nebraska. Its southernmost part extends into northern Phillips County, Kansas. The reservoir is formed by a dam in the Republican River, which starts in Colorado and ends in Kansas.

The two closest towns are Republican City and Alma. Republican City offers fishing supplies, boat rentals, restaurants and cabins. When the lake is at normal pool, Alma is on the shore line of the lake and offers lodging, restaurants, stores, and churches.

History

A number of Native American tribes lived in the Republican River Valley throughout history. The Pawnee, Arapahoe and Cheyenne relied on the land for their survival. Early settlers and trappers arrived in the 1850s when the Nebraska Territory was formed. The rich river bottom soil lured many farmers into the area. According to the U.S. Army Corps of Engineers, Native Americans told early settlers not to build big houses close to the river because of the "big water." In 1872, Dr. John McPherson moved from his Brownville, Nebraska home to the river valley. McPherson opened several stores and laid out lots in a town he called Republican. Once a post office was built, the town took the official name of Republican City. The Burlington Railroad reached the town in 1881, creating a booming trade center.

On May 31, 1935 a storm dumped 24 inches (610 mm) of rain on the Republican River watershed, which averaged nine inches (229 mm) annually. The flood wiped out Republican City within minutes, and 100 people lost their lives. Nearly 10,000 cattle were reportedly lost along the Republican River Valley along with 300 county bridges and over 300 miles (480 km) of roads and railway. The flood was responsible for over $26 million in damages.

The town moved to avoid further damage, and the U.S. Army Corps of Engineers researched a potential location for the dam would work the best to prevent such flood loss. They chose an area between Republican City and Naponee. Republican City was rebuilt two miles (3 km) north on higher ground in 1952, with a few original buildings. Today when the river's water level is low enough you can see some of the original building foundations.

Geography

With a surface covering 13250 acres (54 km^2), the Harlan County Reservoir lake sits in bottom lands of the Republican River Valley. Located seven miles (11 km) from the Nebraska/Kansas border, the lake is nine miles (14 km) long and has about 75 miles (121 km) of shore line. The Reservoir's dam sits on the Republican River two miles (3 km) south of Republican City. According to the U.S. Army Corps of Engineers, the river valley got its name from the Republican Pawnee Indian tribe who inhabited the area. The land consists of thick trees and brush as most river bottoms do. Harlan County Reservoir is Nebraska's second largest lake.

The route of the Republican River

Ecology

The lake is under managements of the U.S. Army Corps of Engineers. The Corps also manages 17750 acres (72 km^2) surrounding the lake as well as the Republican River near the lake. Corps staff conduct safety inspections on the dam, working directly with the Kansas District Water Management to determine how much water is released from the reservoir into the Republican River. Corp Staff are also responsible for the upkeeps of the roads, utility systems, campsites, beaches, boat ramps, and picnic areas in and around the lake. The Corps also works directly with Nebraska Game and Parks Commission to manage the natural resources of Harlan County Lake.

Fish

The Harlan County Lake has fish species such as walleye, white bass, wipers, catfish, crappie and pike. Ice fishing is popular on the reservoir as well. Revenue from fishing licenses is used to further the education and conservation efforts for lakes across Nebraska. According to the Nebraska Game and Parks Commission, money made from issued fishing permits go into efforts such as the Urban Fishing Program.

Birds

According to the Nebraska Birding Trails website, Harlan County Reservoir is home to numerous bird species such as the common loon, Black-legged Kittiwake, parasitic jaeger, little blue heron, white-faced ibis, as well as black-necked stilt and Sabine's gulls. Bald eagles are also often spotted in the latter fall and winter months, especially at the Western end of the lake near Alma. Many migratory birds rely on Harlan county Reservoir each year as they pass through. With an abundance of water in the lake and river, and the thousands of acres of green wheat and corn stalks, this area is crucial as the thousands of ducks and geese fly south to their mating grounds. In past years, there has been a decrease in food available for the birds. According to the Kansas Department of Wildlife and Parks wasted crop from harvest has been reduced from five percent in 1970 to only two percent today due to advancements in equipment. Much of the revenue from hunting permits sold each year go back to help conserve habitat on public lands to ensure an abundance of such birds.

Mammals

The shoreline and rivers surrounding the Reservoir sustain many mammal species. Beaver and muskrat use the muddy banks of the small creeks and streams in and out of the lake as habitat. This is a prime area for such creatures needing the smaller willows and trees to make their dams and huts. The thick brush of the river bottom ground is some of the state's best whitetail deer hunting habitat. Field mice, moles, ground squirrels, rabbits, raccoons, opossum, skunk, badger, coyote, bobcat and many more rely on the habitat supported by the lake for their survival.

Plants

The main habitat of the Harlan County Lake area is mixed prairie grasses. According to the U.S. Army Corps of Engineers these prairies consist of Big and Little Bluestem, Indian grass, western wheatgrass, buffalo grass, blue gramma and switch grass. Much of the area is covered by trees and brush as well. A strand of Cottonwood, ash, cedar, and maple trees just to name a few are found all along the Republican River and other lowland inlets leading up to the lake. The plants of any ecosystem make up most of the habitat.

Land maintenance

A current issue at Harlan County Lake has been a problem of invasive feral hogs destroying a lot of farmland and wildlife habitat while also increasing the chance of introducing disease to other wildlife and domestic hogs in the area. In February 2008 the Nebraska Game and Parks and U.S. Army Corps of Engineers exterminated 69 feral hogs from the Harlan County Lake area. They were shot from a helicopter by a professional sharpshooter from the Kansas Wildlife Services. A large amount of native prairie and woodland have been converted back over the years in the area. Over 300,000 trees have been planted and 2000 acres (8 km^2) of native prairie have been planted using funds from the Nebraska Game and Parks Commission Habitat Stamp Program.

See also

- List of Nebraska lakes
- List of Nebraska fish
- List of Nebraska rivers
- List of Nebraska streams

References

- Harlen County Lake [1] website.
- "Nebraska Game and Parks Commission Targets Feral Hogs in Harlan County" [2], OutdoorNewsDaily.com.
- Nebraska Wildlife Federation [3] website.
- Harlan County Reservoir [4] webpage on the US Army Corps of Engineers website.
- Harlan County Reservoir [5] information on the Nebraska Bird Trails website.

References

[1] http://www.harlanlake.com/
[2] http://outdoornewsdaily.com/index.php/archives/2914
[3] http://www.nebraskawildlife.org/policy.html
[4] http://www.nwk.usace.army.mil/hc/
[5] http://www.nebraskabirdingtrails.com/site.asp?site=374

Nebraska

State of Nebraska

Flag

Seal

Nickname(s): *Cornhusker State*

Motto(s): *Equality Before the Law*

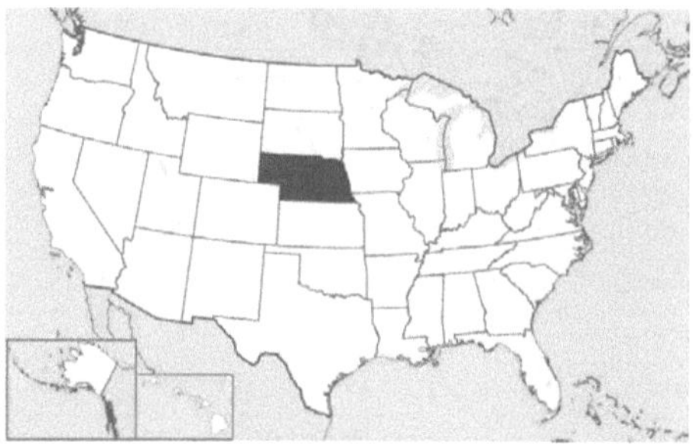

Official language(s)	English
Demonym	Nebraskan
Capital	Lincoln
Largest city	Omaha
Largest metro area	Omaha-Council Bluffs
Area	Ranked 16th in the U.S.
- Total	77,354 sq mi (200,520 km^2)
- Width	210 miles (340 km)
- Length	430 miles (690 km)
- % water	0.7
- Latitude	40° N to 43° N
- Longitude	95° 19' W to 104° 03' W
Population	Ranked 38th in the U.S.
- Total	1,826,341 [1]
- Density	23/sq mi (8.88/km^2) Ranked 43rd in the U.S.
- Median income	$44,623 (20th)

Elevation	
- Highest point	Panorama Point[2] 5,424 ft (1,653 m)
- Mean	2,592 ft (790 m)
- Lowest point	Missouri River[2] 840 ft (256 m)
Admission to Union	March 1, 1867 (37th)
Governor	Dave Heineman (R)
Lieutenant Governor	Rick Sheehy (R)
Legislature	Nebraska Legislature
- Upper house	*None*
- Lower house	*None*
U.S. Senators	Ben Nelson (D) Mike Johanns (R)
U.S. House delegation	Jeff Fortenberry (R) Lee Terry (R) Adrian M. Smith (R) (list)
Time zones	
- most of state	Central: UTC-6/-5
- panhandle	Mountain: UTC-7/-6
Abbreviations	NE US-NE
Website	[www.nebraska.gov www.nebraska.gov]

Nebraska i/nəˈbræskə/ is a state on the Great Plains of the Midwestern United States. The state's capital is Lincoln and its largest city is Omaha, on the Missouri River.

Once considered part of the Great American Desert, Nebraska is now a leading farming and ranching state. The Omaha and Ho-Chunk (Winnebago) are federally recognized American Indian tribes, each with a reservation in Thurston County in eastern Nebraska, where they constitute the majority of population.

Etymology

Nebraska gets its name from the archaic Otoe words *Ñí Brásge*, pronounced [ɲĩbɾaskɛ] (contemporary Otoe *Ñí Bráhge*), or the Omaha *Ní Btháska*, pronounced [nĩbɫ aska], meaning "flat water", after the Platte River that flows through the state.[3]

History

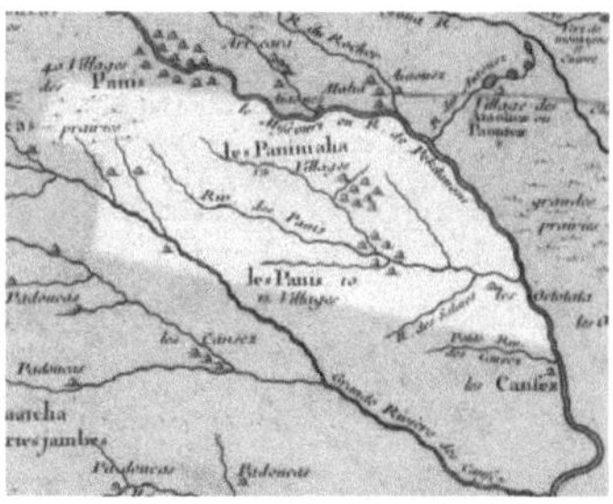

Nebraska in 1718, Guillaume de L'Isle map, with the approximate area of the future state highlighted

Varying cultures of indigenous peoples lived in the region along the rivers for thousands of years before European exploration. Historical Native American tribes living in Nebraska have included the Omaha, Missouria, Ponca, Pawnee, Otoe, and various branches of the Lakota (Sioux).

Long before the Lewis and Clark Expedition of 1804-1806, French-Canadian explorers (including the Mallet brothers in 1739) traversed the territory of Nebraska on their way to trade in Santa Fe, then claimed by Spain.[4]

In 1819, the United States established Fort Atkinson as the first US Army post west of the Missouri River, just east of present-day Fort Calhoun. The army abandoned the fort in 1827 as migration moved further west.

European-American settlement did not begin in any numbers until after 1848 and the California Gold Rush. On May 30, 1854, the US Congress created the Kansas and the Nebraska territories, divided by the Parallel 40° North, under the Kansas-Nebraska Act.[5] The Nebraska Territory included parts of the current states of Colorado, North Dakota, South Dakota, Wyoming, and Montana.[6] The territorial capital of Nebraska was Omaha.

In the 1860s, after the US government forced many of the Native American tribes to cede their lands and settle on reservations, it opened large tracts of land to agricultural development by Europeans and Americans. Under the Homestead Act, thousands of new settlers migrated into Nebraska to claim free land granted by the federal government. Because so few trees grew on the prairies, many of the first farming settlers built their homes of sod, as had the Native Americans such as the Omaha. The first wave of settlement gave the territory a sufficient population to apply for statehood.[7]

Nebraska became the 37th state on March 1, 1867, approximately two years after the American Civil War. At that time, the capital was moved from Omaha to the center at Lancaster, later renamed Lincoln after the recently assassinated President of the United States, Abraham Lincoln.

During the 1870s to the 1880s, Nebraska experienced a large growth in population. Several factors contributed to attracting new residents. The first was that the vast prairie land was perfect for cattle grazing. This helped settlers to learn the unfamiliar geography of the area. The second factor was the invention of several farming technologies. Agricultural inventions such as barbed wire, wind mills, and the steel plow, combined with good weather, enabled settlers to make use of Nebraska as prime farming land. By the 1880s, Nebraska's population had soared to more than 450,000 people.[8]

The Arbor Day holiday was founded in Nebraska. The National Arbor Day Foundation is still headquartered in Nebraska City, with some offices in Lincoln.

In the late nineteenth century, many African Americans migrated from the South to Nebraska as part of the Great Migration, primarily to Omaha which offfered working class jobs in meatpacking, the railroads and other industries. Omaha has a long history of civil rights activism. Blacks encountered discrimination from other Americans in Omaha and especially from recent European immigrants, ethnic whites who were competing for the same jobs. In

1912 African Americans founded the Omaha chapter of the National Association for the Advancement of Colored People to work for improved conditions in the city and state. Activism has continued.

Since the 1960s, Native American activism in the state has increased, both through open protest, activities to build alliances with state and local governments, and in the slower, more extensive work of building tribal institutions and infrastructure. Native Americans in federally recognized tribes have pressed for self-determination, sovereignty and recognition. They have created community schools to preserve their cultures, as well as tribal colleges and universities. Tribal politicians have also collaborated with state and county officials on regional issues.

Geography

Further information: List of counties in Nebraska, List of Nebraska rivers, and Geography of Omaha

The state is bordered by South Dakota to the north; Iowa to the east and Missouri to the southeast, across the Missouri River; Kansas to the south; Colorado to the southwest; and Wyoming to the west. The state has 93 counties; it occupies the central portion of the Frontier Strip. Nebraska is split into two time zones. The Central Time zone comprises the eastern half of the state, while the western half observes Mountain Time. Three rivers cross the state from west to east. The Platte River, formed by the confluence of the North Platte and the South Platte, runs through the central portion of the state, the Niobrara River flows through the northern part, and the Republican River runs across the southern part.

Scotts Bluff National Monument

Nebraska is composed of two major land regions: the Dissected Till Plains and the Great Plains. The easternmost portion of the state was scoured by Ice Age glaciers; the Dissected Till Plains were left behind after the glaciers retreated. The Dissected Till Plains is a region of gently rolling hills; Omaha and Lincoln are in this region.

The Great Plains occupy the majority of western Nebraska. The Great Plains region consists of several smaller, diverse land regions, including the Sandhills, the Pine Ridge, the Rainwater Basin, the High Plains and the Wildcat Hills. Panorama Point, at 5,424 feet (1,653 m), is the highest point in Nebraska; despite its name and elevation, it is a relatively low rise near the Colorado and Wyoming borders.

A past Nebraska tourism slogan was "Where the West Begins"; locations given for the beginning of the "West" include the Missouri River, the intersection of 13th and O Streets in Lincoln (where it is marked by a red brick star), the 100th meridian, and Chimney Rock. Nebraska is a triply landlocked state, as it does not border the ocean, nor do any of the states it borders, nor any that they border.[9]

Nebraska state welcome sign on Interstate 76 at the Colorado border

Federal land management

Areas under the management of the National Park Service include:

- Agate Fossil Beds National Monument near Harrison
- California National Historic Trail

- Chimney Rock National Historic Site near Bayard
- Homestead National Monument of America in Beatrice
- Lewis & Clark National Historic Trail
- Missouri National Recreational River near Ponca
- Mormon Pioneer National Historic Trail
- Niobrara National Scenic River near Valentine
- Oregon National Historic Trail
- Pony Express National Historic Trail
- Scotts Bluff National Monument at Gering

Areas under the management of the National Forest Service include:

- Nebraska National Forest
- Oglala National Grassland
- Samuel R. McKelvie National Forest

Climate

Two major climates are represented in Nebraska: the eastern half of the state has a humid continental climate (Köppen climate classification *Dfa*), and the western half of the state has a semi-arid climate (Koppen *BSk*). The entire state experiences wide seasonal variations in temperature and precipitation. Average temperatures are fairly uniform across Nebraska with hot summers and generally cold winters, while average annual precipitation decreases east to west from about 31.5 inches (800 mm) in the southeast corner of the state to about 13.8 inches (350 mm) in the Panhandle. Humidity also decreases significantly from east to west. Snowfall across the state is fairly even, with most of Nebraska receiving between 25 and 35 inches (65 to 90 cm) of snow annually.[10] Nebraska's highest recorded temperature is 118 °F (48 °C) at Minden on July 24, 1936 and the lowest recorded temperature is −47 °F (−44 °C) at Camp Clarke on February 12, 1899.

Nebraska is in Tornado Alley; thunderstorms are common in the spring and summer months, and violent thunderstorms and tornadoes happen primarily during the spring and summer, though they can also happen in the autumn. The chinook winds from the Rocky Mountains provide a temporary moderating effect on temperatures in western Nebraska during the winter months.[11] [12]

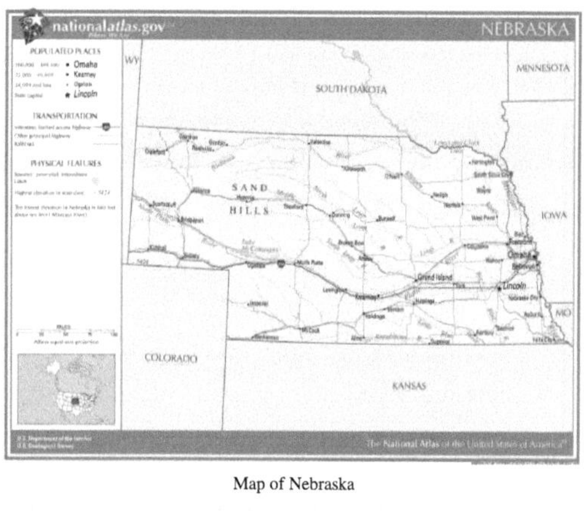

Map of Nebraska

Demographics

Historical populations		
Census	Pop.	%±
1860	28841	—
1870	122993	326.5%
1880	452402	267.8%
1890	1062656	134.9%
1900	1066300	0.3%
1910	1192214	11.8%
1920	1296372	8.7%
1930	1377963	6.3%
1940	1315834	−4.5%
1950	1325510	0.7%
1960	1411330	6.5%
1970	1483493	5.1%
1980	1569825	5.8%
1990	1578385	0.5%
2000	1711263	8.4%
2010	1826341	6.7%
Source: 1910-2010[13]		

As of 2009, Nebraska has an estimated population of 1,796,619, which is an increase of 85,356, or 5%, since the year 2000. This includes a natural increase since the last census of 77,995 people (that is 187,564 births minus 109,569 deaths) and a decrease due to net migration of 9,319 people out of the state.

The center of population of Nebraska is in Polk County, in the city of Shelby.[14]

As of 2004, the population of Nebraska included about 84,000 foreign-born residents (4.8% of the population).

Demographics of Nebraska (csv) [15]

By race	White	Black	AIAN*	Asian	NHPI*
2000 (total population)	93.53%	4.48%	1.32%	1.58%	0.11%
2000 (Hispanic only)	5.24%	0.13%	0.17%	0.04%	0.02%
2005 (total population)	93.06%	4.82%	1.31%	1.85%	0.11%
2005 (Hispanic only)	6.74%	0.20%	0.18%	0.05%	0.03%
Growth 2000–05 (total population)	2.25%	10.52%	1.72%	20.76%	5.53%
Growth 2000–05 (non-Hispanic only)	0.48%	9.02%	0.56%	20.73%	4.75%
Growth 2000–05 (Hispanic only)	32.17%	61.17%	9.63%	22.13%	8.10%
* AIAN is American Indian or Alaskan Native; NHPI is Native Hawaiian or Pacific Islander					

The five largest ancestry groups in Nebraska are German (38.6%), Irish (12.4%), English (9.6%), Swedish (4.9%), and Czech (4.9%).

Nebraska has the largest Czech-American and non-Mormon Danish-American population (as a percentage of the total population) in the nation. German Americans are the largest ancestry group in most of the state, particularly in the eastern counties. Thurston County (made up entirely of the Omaha and Winnebago reservations) has an American Indian majority, and Butler County is one of only two counties in the nation with a Czech-American plurality.

Rural flight

Eighty-nine percent of the cities in Nebraska have fewer than 3,000 people. Nebraska shares this characteristic with five other Midwestern states: Kansas, Oklahoma, North and South Dakota, and Iowa. Hundreds of towns have a population of fewer than 1,000. Regional population declines have forced many rural schools to consolidate.

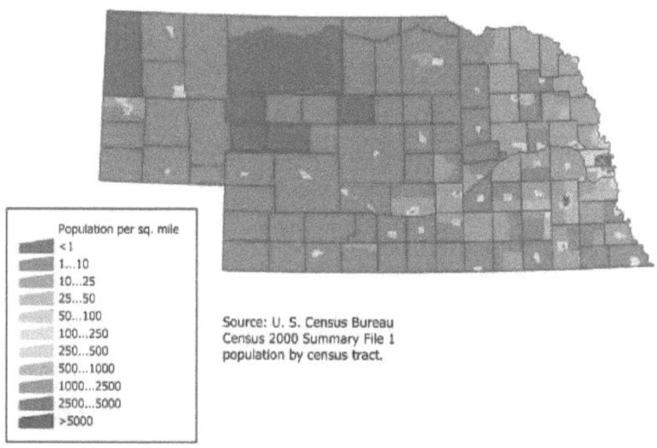

Population density of Nebraska

Fifty-three of Nebraska's 93 counties reported declining populations between 1990 and 2000, ranging from a 0.06% loss (Frontier County) to a 17.04% loss (Hitchcock County).

More urbanized areas of the state have experienced substantial growth. In 2000, the city of Omaha had a population of 390,007; in 2005, the city's estimated population was 414,521 (427,872 including the recently annexed city of Elkhorn), a 6.3% increase over five years. The 2010 census showed that Omaha has a population of 408,958. The city of Lincoln had a 2000 population of 225,581 and a 2010 population of 258,379, a 14.5% increase.

Religion

The religious affiliations of the people of Nebraska are:

- Christian – 90%
  - Catholic – 28%
  - Lutheran – 16%
  - Methodist – 11%
  - Baptist – 9%
  - Presbyterian – 4%
    - Other Protestant – 21%
  - Other Christian – 1%
- Non-religious – 9%
- Other religions – 1%

The largest single denominations by number of adherents in 2000 were the Catholic Church (372,791), the Evangelical Lutheran Church in America (128,570), the Lutheran Church - Missouri Synod (117,419) and the United Methodist Church (117,277).[16]

Taxation

Nebraska has a progressive income tax. The portion of income from $0 to $2,400 is taxed at 2.56%; from $2,400 to $17,500, at 3.57%; from $17,500 to $27,000, at 5.12%; and income over $27,000, at 6.84%. The standard deduction for a single taxpayer is $5,700; the personal exemption is $118.[17]

Nebraska has a state sales tax of 5.5%. In addition to the state tax, some Nebraska cities assess a city sales and use tax, up to a maximum of 1.5%. One county in Nebraska, Dakota County, levies a sales tax. All real property within the state of Nebraska is taxable unless specifically exempted by statute. Since 1992, only depreciable personal property is subject to tax and all other personal property is exempt from tax. Inheritance tax is collected at the county level.

Economy

The Bureau of Economic Analysis estimates of Nebraska's gross state product in 2010 was $89.8 billion.[18] Per capita personal income in 2004 was $31,339, 25th in the nation. Nebraska has a large agriculture sector, and is an important producer of beef, pork, corn (maize), and soybeans.[19] Other important economic sectors include freight transport (by rail and truck), manufacturing, telecommunications, information technology, and insurance.

As of January 2010, the state's unemployment rate is 4.6%.[20]

Industry

Kool-Aid was created in 1927 by Edwin Perkins in the city of Hastings, which celebrates the event the second weekend of every August with Kool-Aid Days [21]. Kool-Aid is the official soft drink of Nebraska.[22] *CliffsNotes* were developed by Clifton Hillegass of Rising City. He adapted his pamphlets from the Canadian publications, *Coles Notes*.

Omaha is home to Berkshire Hathaway, whose CEO Warren Buffett was ranked in March 2009 by *Forbes* magazine as the second richest person in the world. The city is also home to ConAgra, Mutual of Omaha, InfoUSA, TD Ameritrade, West Corporation, Valmont Industries, Woodmen of the World, Kiewit Corporation, and the Union Pacific Railroad. UNIFI Companies, Sandhills Publishing Company, and Duncan Aviation are based in Lincoln; The Buckle is based in Kearney. Sidney is the national headquarters for Cabela's, a specialty retailer of outdoor goods.

The world's largest train yard, Union Pacific's Bailey Yard, is in North Platte. The Vise-Grip was invented by William Petersen in 1924, and was manufactured in De Witt until the plant was closed and moved to China in late 2008.[23]

Lincoln's Kawasaki Motors Manufacturing is the only Kawasaki plant in the world to produce the Jet-Ski, ATV, and Mule lines of product. The facility employs more than 1200 people.

The Spade Ranch, in the Sand Hills, is one of Nebraska's oldest and largest beef cattle operations.

Transportation

Railroads

Further information: List of Nebraska railroads

Nebraska has a rich railroad history. The Union Pacific Railroad, headquartered in Omaha, was incorporated on July 1, 1862, in the wake of the Pacific Railway Act of 1862. Bailey Yard, in North Platte, is the largest railroad classification yard in the world. The route of the original transcontinental railroad runs through the state.

Other major railroads with operations in the state are: Amtrak; Burlington Northern & Santa Fe Railway; Canadian Pacific Railway; and Iowa Interstate Railroad.

Roads and highways

Further information: List of Nebraska numbered highways

Interstate Highways through the State of Nebraska

The U.S. Routes in Nebraska

Law and government

Nebraska's government operates under the framework of the Nebraska Constitution, adopted in 1875,[24] and is divided into three branches: executive, legislative, and judicial.

Executive branch

Further information: Governor of Nebraska

The head of the executive branch is the Governor Dave Heineman. Other elected officials in the executive branch are the Lieutenant Governor Rick Sheehy (elected on the same ticket as the Governor), Attorney General Jon Bruning, Secretary of State John A. Gale, State Treasurer Shane Osborn, and State Auditor Mike Foley. All elected officials in the executive branch serve four-year terms.

Legislative branch

Further information: Nebraska Legislature and Nebraska State Capitol

Nebraska is the only state in the United States with a unicameral legislature. Although this house is officially known simply as the "Legislature", and more commonly called the "Unicameral", its members call themselves "senators". Nebraska's Legislature is also the only state legislature in the United States that is nonpartisan. The senators are elected with no party affiliation next to their names on the ballot, and the speaker and committee chairs are chosen at large, so that members of any party can be chosen for these positions. The Nebraska Legislature can also override a governor's veto with a three-fifths majority, in contrast to the two-thirds majority required in some other states.

The Nebraska Legislature meets in the third Nebraska State Capitol building, built between 1922 and 1932. It was designed by Bertram G. Goodhue. Built from Indiana limestone, the Capitol's base is a cross within a square. A 400-foot domed tower rises from this base. The Golden Sower, a 19-foot bronze statue representing agriculture, crowns the Capitol. The state Capitol is considered an architectural achievement and has been recognized by the American Institute of Architects.

Nebraska state insignia	
Motto	*Equality Before the Law*
Slogan	*Nebraska, possibilities...endless*
Bird	Western meadowlark
Animal	White-tailed deer
Fish	Channel catfish
Insect	European honey bee
Flower	Goldenrod
Tree	Cottonwood
Song	"Beautiful Nebraska"
Quarter	 *Released April 7, 2006*
Grass	Little bluestem
Beverage	Milk
Dance	Square dance
Fossil	Mammoth
Gemstone	Blue agate
Rock	Prairie agate
Soil	Holdrege series

For years, United States Senator George Norris and other Nebraskans encouraged the idea of a unicameral legislature, and demanded the issue be decided in a referendum. Norris argued:

> "The constitutions of our various states are built upon the idea that there is but one class. If this be true, there is no sense or reason in having the same thing done twice, especially if it is to be done by two bodies of men elected in the same way and having the same jurisdiction."

Unicameral supporters also argued that a bicameral legislature had a significant undemocratic feature in the committees that reconciled Assembly and Senate legislation. Votes in these committees were secretive, and would sometimes add provisions to bills that neither house had approved. Nebraska's unicameral legislature today has rules that bills can contain only one subject, and must be given at least five days of consideration.

In 1934, due in part to the budgetary pressure of the Great Depression, Nebraska citizens ran a state initiative to vote on a unicameral legislature, which was approved. In effect, the Assembly (the house) was abolished; as noted, today's Nebraska state legislators are commonly referred to as "Senators".

Judicial branch

Further information: Nebraska Supreme Court

The judicial system in Nebraska is unified, with the Nebraska Supreme Court having administrative authority over all Nebraska courts. Nebraska uses the Missouri Plan for the selection of judges at all levels. The lowest courts in Nebraska are the county courts, above that are twelve district courts (containing one or more counties). The Court of Appeals hears appeals from the district courts, juvenile courts, and workers' compensation courts. The Nebraska Supreme Court is the final court of appeal.

In 2008, the Nebraska Supreme Court ruled that the state's only method of execution, electrocution, was in conflict with the state's constitution. For the next year, Nebraska had no active death-penalty law. (Prior to that ruling, Nebraska was the only place in the world that used electrocution as the sole method of execution.) In May 2009, the legislature passed and the governor signed a bill that changed the method of execution in Nebraska to lethal injection, enabling capital punishment.[25] Executions in Nebraska have been infrequent; none have been carried out in the 21st century. During the last few decades, residents have considered a moratorium on, or complete abolition of, capital punishment.

Federal government representation

Further information: United States congressional delegations from Nebraska

The Nebraska State Capitol in Lincoln, Nebraska

Nebraska's U.S. senators are Mike Johanns (R), the junior senator, and Ben Nelson (D), the senior senator.

Nebraska has three representatives in the House of Representatives: Jeff Fortenberry (R) of the 1st district; Lee Terry (R) of the 2nd district; and Adrian M. Smith (R) of the 3rd district.

Nebraska is one of two states (with Maine) that allow for a split in the state's allocation of electoral votes in presidential elections. Under a 1991 law, two of Nebraska's five votes are awarded to the winner of the statewide popular vote, while the other three go to the highest vote-getter in each of the state's three congressional districts.

In the 2008 election, the Democrat Barack Obama won the 2nd congressional district (which includes Omaha), giving him the electoral vote for that district.[26] The Republican John McCain won the statewide popular vote, with 58% of the vote to Obama's 42%; this, combined with his victories in the 1st and 3rd districts, gave him Nebraska's other four electoral votes.[27]

Politics

Further information: United States presidential election, 2004, in Nebraska, Nebraska gubernatorial election, 2006, Nebraska United States Senate election, 2006 and Political party strength in Nebraska

For most of its history, Nebraska has been a solidly Republican state. Republicans have carried the state in all but one presidential election since 1940—the 1964 landslide election of Lyndon B. Johnson. In the 2004 presidential election, George W. Bush won the state's five electoral votes by a 33% margin (the fourth-most Republican vote among states) with 65.9% of the overall vote; only Thurston County, which is majority-Native American, voted for John Kerry.

Despite the current Republican domination of Nebraska politics, the state has a long tradition of electing centrist members of both parties to state and federal office; examples include George Norris (who served few years in the Senate as an independent), J. James Exon, and Bob Kerrey. Voters have tilted to the right in recent years with the election of conservative Mike Johanns to the US Senate and the re-election of Ben Nelson, who is currently considered the most conservative Democrat in the US Senate.

Important cities and towns

All population figures are 2010 Census Bureau estimates.

Largest cities

- Omaha – 408,958
- Lincoln – 258,379
- Bellevue – 50,137
- Grand Island – 48,520
- Kearney – 30,787
- Fremont – 26,397
- Hastings – 24,907
- North Platte – 24,733
- Norfolk – 24,210
- Columbus – 22,111
- Papillion – 18,894
- La Vista – 15,758
- Scottsbluff – 15,039
- South Sioux City – 13,353
- Beatrice – 12,459
- Lexington – 10,230
- Gering - 8,500
- Alliance - 8,491
- Blair - 7,990
- York - 7,766

Downtown Omaha

Downtown Lincoln

Urban areas

Metropolitan areas

- Omaha-Council Bluffs – 728,963 (Nebraska portion); 849,517 (total for Nebraska and Iowa)
- Lincoln – 298,012
- Sioux City, Iowa – 26,940 (Nebraska portion); 144,360 (total for Nebraska, Iowa and South Dakota)

Micropolitan areas

- Grand Island – 71,596
- Kearney – 52,274
- Norfolk – 48,000
- Hastings – 39,529
- Scottsbluff – 37,512
- North Platte – 36,890
- Fremont – 35,640
- Columbus – 32,515
- Lexington – 26,920
- Beatrice – 22,653

Other areas

- Grand Island, Hastings and Kearney comprise the "Tri-Cities" area, with a combined population of 163,399.
- The northeast corner of Nebraska is part of the Siouxland region.

Education

Colleges and universities

University of Nebraska system

- University of Nebraska–Lincoln
- University of Nebraska at Kearney
- University of Nebraska at Omaha
- University of Nebraska Medical Center
- Nebraska College of Technical Agriculture

Nebraska State College System

- Chadron State College
- Peru State College
- Wayne State College

Nebraska Community College Association

- Central Community College
- Little Priest Tribal College
- Metropolitan Community College
- Mid-Plains Community College
- Nebraska Indian Community College
- Northeast Community College
- Southeast Community College
- Western Nebraska Community College

Private colleges/universities

- Bellevue University
- Clarkson College
- College of Saint Mary
- Concordia University
- Creighton University
- Doane College
- Grace University
- Hastings College
- Midland University
- Nebraska Christian College
- Nebraska Methodist College
- Nebraska Wesleyan University
- Summit Christian College
- Union College
- York College

Further information: Colleges and universities of Omaha, Nebraska

Culture

- Arbor Day was founded by J. Sterling Morton; the National Arbor Day Foundation has its headquarters near his home in Nebraska City.
- The swing in the Hebron, Nebraska city park at 5th and Jefferson streets is claimed to be the world's largest porch swing, long enough to fit 18 adults or 26 children.
- Nebraska Huskers football influences many of Nebraska's residents. During home football games, Memorial Stadium in Lincoln, with a capacity of 85,500, becomes larger than Nebraska's third-largest city.[28]
- Job's Daughters was founded in Omaha in 1920 by Ethel T. Wead Mick. There are now bethels in Canada, Australia, Brazil, and Philippines.

Sports

Further information: Sports in Nebraska

- Professional sports
  - Lincoln Saltdogs – American Association (independent minor league baseball)
  - Nebraska Danger - Indoor Football League
  - Omaha Beef – Indoor Football League
  - Omaha Storm Chasers – Pacific Coast League (AAA minor league baseball; affiliate of the Kansas City Royals)
  - Omaha Nighthawks – United Football League
  - Omaha Vipers - Major Indoor Soccer League
- NCAA Division I college sports
  - Creighton Bluejays
  - Nebraska Cornhuskers
  - University of Nebraska at Omaha (currently ice hockey only; all other sports moving to Division I in fall 2012)
- NCAA Division II college sports[29]
  - University of Nebraska at Kearney Antelopes (Lopers)

- University of Nebraska at Omaha Mavericks
- Wayne State College Wildcats
- Chadron State College Eagles
- Junior-level sports: United States Hockey League
    - Lincoln Stars
    - Omaha Lancers
    - Tri-City Storm

See also

- List of National Register of Historic Places in Nebraska

References

[1] "Annual Estimates of the Resident Population for the United States, Regions, States, and Puerto Rico: April 1, 2000 to July 1, 2008" (http://www.census.gov/popest/states/tables/NST-EST2008-01.csv). United States Census Bureau. . Retrieved 2009-01-26.

[2] "Elevations and Distances in the United States" (http://erg.usgs.gov/isb/pubs/booklets/elvadist/elvadist.html#Highest). U.S Geological Survey. 29 April 2005. . Retrieved November 6, 2006.

[3] Koontz, John. "Etymology" (http://spot.colorado.edu/~koontz/faq/etymology.htm). *Siouan Languages*. . Retrieved 2006-11-28.

[4] "Nebraska", *Catholic Encyclopedia*, 1910

[5] NebraskaStudies.org. 2009. "1854: Kansas-Nebraska Act Signed." http://nebraskastudies.unl.edu/0500/frameset_reset.html?http://nebraskastudies.unl.edu/0500/stories/0502_0100.html . Retrieved February 3, 2009.

[6] *The Handybook for Genealogists: United States of America*, 10th ed. (Draper Utah: Everton Publishers, 2002).

[7] Marsha Hoffman and Dwight A. Radford, "Nebraska," *Redbook: American State, County, and Town Sources*, 3rd ed. (Provo: Ancestry, 2004), 408.

[8] *Redbook*

[9] Note that Nebraska's designation as a "triply landlocked" state does not consider the fact that Minnesota, Wisconsin and Illinois have ports on the shores of the Great Lakes that have access to the seas.

[10] http://www.hprcc.unl.edu/products/atlas.html

[11] "Nebraska Climate Office I Applied Climate Science I SNR I UNL" (http://www.nebraskaclimateoffice.unl.edu/weather.summary.Dec99). Nebraskaclimateoffice.unl.edu. 2009-07-23. . Retrieved 2010-04-17.

[12] "Climate – Twin Cities Development Association, Inc. – Nebraska: Scottsbluff, Gering, TerryTown, Mitchell, Bayard" (http://www.tcdne.org/climate.htm). Tcdne.org. . Retrieved 2009-02-24.

[13] http://2010.census.gov/2010census/data/apportionment-pop-text.php

[14] "Population and Population Centers by State: 2000" (http://www.census.gov/geo/www/cenpop/statecenters.txt). United States Census Bureau. . Retrieved 2008-12-05.

[15] http://www.census.gov/popest/states/asrh/tables/SC-EST2005-03-31.csv

[16] "State Membership Report". (http://www.thearda.com/mapsReports/reports/state/31_2000.asp) *The Association of Religion Data Archives*. Retrieved July 29, 2008.

[17] "State Individual Income Tax Rates, 2000–2010" (http://www.taxfoundation.org/publications/show/228.html). The Tax Foundation. 2010-03-25. . Retrieved 2011-03-03.

[18] "GDP by State" (http://greyhill.com/gdp-by-state). Greyhill Advisors. . Retrieved 7 September 2011.

[19] "Nebraska State Agriculture Overview – 2006" (http://www.nass.usda.gov/Statistics_by_State/Ag_Overview/AgOverview_NE.pdf) (PDF). United States Department of Agriculture. . Retrieved 2007-10-17.

[20] Bls.gov (http://www.bls.gov/lau/); Local Area Unemployment Statistics

[21] http://www.kool-aiddays.com

[22] "History: Kool-Aid: Hastings Museum" (http://www.hastingsmuseum.org/koolaid/kahistory.htm). Hastings Museum<!. . Retrieved 2009-02-24.

[23] Jirovsky, Kristin. "Owner of Nail Jack Tools wants to share former Vise-Grip plant" (http://journalstar.com/articles/2009/01/08/news/business/doc4966307080dcd635956810.txt), *Lincoln Journal-Star Online*. Jan 8, 2009.

[24] "Nebraska as a State". (http://www.kancoll.org/books/andreas_ne/state/state-p6.html) *Andreas's History of the State of Nebraska*. (http://www.kancoll.org/books/andreas_ne/hon_tabl.html). Retrieved 2010-02-18.

[25] Volentine, Jason (2009-05-28). "Nebraska Changes Execution Method to Lethal Injection" (http://www.kolnkgin.com/political/headlines/46424547.html). KOLN. . Retrieved 2009-05-30.

[26] Ortiz, Jean. "Bill targets Nebraska's ability to split electoral votes" (http://journalstar.com/news/local/govt-and-politics/article_b05a487c-fbae-11de-b9c2-001cc4c002e0.html), *Lincoln Journal Star*, 7 January 2010, Retrieved 2010-10-24.

[27] Walton, Don. "Omaha's electoral vote still undetermined" (http://journalstar.com/news/local/govt-and-politics/article_23cc1f78-3d8e-58de-8644-8cbcbfbf6046.html), *Lincoln Journal Star*, 4 November 2008, Retrieved 2010-10-24

[28] "Nebraska Lost, Nebraska Found" (http://vault.sportsillustrated.cnn.com/vault/article/magazine/MAG1128181/index.htm). *Sports Illustrated*. 2008-04-21. . Retrieved 2008-05-01.

[29] "NCAA Division II Home Page" (http://web.archive.org/web/20070822001230/http://www.siue.edu/ATHLETIC/d2/alpha.html). National Collegiate Athletic Association. Archived from the original (http://www.siue.edu/ATHLETIC/d2/alpha.html) on 2007-08-22. . Retrieved 2007-08-30.

Bibliography

Surveys

- Chokecherry Places, Essays from the High Plains, Merrill Gilfillan, Johnson Press, Boulder, Colorado, trade paperback, ISBN 1-55566-227-7.
- Olson James C. and Ronald C. Naugle, *History of Nebraska* (http://www.questia.com/PM.qst?a=o&d=28579910) 2nd ed (1997)
- Andreas, Alfred T., *History of the State of Nebraska* (http://www.kancoll.org/books/andreas_ne/) (1882) (a highly detailed history)
- Creigh, Dorothy Weyers. *Nebraska: A Bicentennial History* (1977)
- Faulkner, Virginia, ed. *Roundup: A Nebraska Reader* (1957)
- Hickey, Donald R. *Nebraska Moments: Glimpses of Nebraska's Past* (1992).
- Miewald, Robert D., *Nebraska Government & Politics* (http://www.questia.com/PM.qst?a=o&d=51597481) (1984)
- Luebke Frederick C. *Nebraska: An Illustrated History* (1995)
- Morton, J. Sterling, ed. *Illustrated History of Nebraska: A History of Nebraska from the Earliest Explorations of the Trans-Mississippi Region.* 3 vols. (1905–13)
- Wishart, David J. ed. *Encyclopedia of the Great Plains* (2004), 900 pages of scholarly articles
- *Nebraska: A Guide to the Cornhusker State*, WPA Guide, 1939; Online full-text PDF edition (http://digitalcommons.unl.edu/cgi/siteview.cgi//nebraskianapubs/2)

Scholarly special studies

- Barnhart, John D. "Rainfall and the Populist Party in Nebraska." *American Political Science Review* 19 (1925): 527–40. in JSTOR
- Beezley, William H. "Homesteading in Nebraska, 1862–1872", *Nebraska History* 53 (spring 1972): 59–75
- Bentley, Arthur F. "The Condition of the Western Farmer as Illustrated by the Economic History of a Nebraska Township." *Johns Hopkins University Studies in Historical and Political Science* 11 (1893): 285–370
- Cherny, Robert W. *Populism, Progressivism, and the Transformation of Nebraska Politics, 1885–1915* (1981) (http://www.questia.com/PM.qst?a=o&d=89553965)
- Bogue Allen G. *Money at Interest: The Farm Mortgage on the Middle Border* (1955)
- Brunner, Edmund de S. *Immigrant Farmers and Their Children* (1929)
- Chudacoff, Howard P. *Mobile Americans: Residential and Social Mobility in Omaha, 1880–1920* (1972)
  - Chudacoff, Howard P. "A New Look at Ethnic Neighborhoods: Residential Dispersion and the Concept of Visibility in a Medium-sized City." *Journal of American History* 60 (1973): 76–93. about Omaha; in JSTOR
- Coletta, Paolo E. *William Jennings Bryan*. (http://www.questia.com/PM.qst?a=o&d=105476144) 3 vols. (1964–69)
- Dick, Everett. *The Sod-House Frontier: 1854–1890* (1937)
- Farragher, John Mack. *Women and Men on the Overland Trail* (1979)
- Fuller, Wayne E. *The Old Country School: The Story of Rural Education in the Midwest* (1982)
- Grant, Michael Johnston. "Down and Out on the Family Farm" (2002)

- Harper, Ivy. *Walzing Matilda: Life and Times of Nebraska Senator Robert Kerrey* (1992)
- Holter, Don W. *Flames on the Plains: A History of United Methodism in Nebraska* (1983)
- Jeffrey, Julie Roy. *Frontier Women: The Trans-Mississippi West, 1840–1880* (1979)
- Klein, Maury. *Union Pacific: The Birth of a Railroad, 1862–1893* (1986)
- Klein, Maury. *Union Pacific: The Rebirth, 1894–1969* (1989)
- Larsen, Lawrence H. *The Gate City: A History of Omaha* (1982)
- Lowitt, Richard. *George W. Norris* 3 vols. (1971)
- Luebke, Frederick C. *Immigrants and Politics: The Germans of Nebraska, 1880–1900* (1969)
- Luebke, Frederick C. "The German-American Alliance in Nebraska, 1910–1917." *Nebraska History* 49 (1969): 165–85
- Olson, James C. *J. Sterling Morton* (1942)
- Overton, Richard C. *Burlington West: A Colonization History of the Burlington Railroad* (1941)
- Parsons Stanley B. "Who Were the Nebraska Populists?" *Nebraska History* 44 (1963): 83–99
- Pierce, Neal. *The Great Plains States* (1973)
- Pederson, James F., and Kenneth D. Wald. *Shall the People Rule? A History of the Democratic Party in Nebraska Politics* (1972)
- Riley, Glenda. *The Female Frontier. A Comparative View of Women on the Prairie and the Plains* (1978)
- Wenger, Robert W. "The Anti-Saloon League in Nebraska Politics, 1898–1910." *Nebraska History* 52 (1971): 267–92

External links

- Nebraska state government (http://www.nebraska.gov/)
- Nebraska Division of Travel and Tourism (http://www.visitnebraska.gov/)
- Energy Profile for Nebraska (http://tonto.eia.doe.gov/state/state_energy_profiles.cfm?sid=NE)
- USGS real-time, geographic, and other scientific resources of Nebraska (http://www.usgs.gov/state/state.asp?State=NE)
- U.S. Census Bureau (http://quickfacts.census.gov/qfd/states/31000.html)
- Nebraska State Facts (http://www.ers.usda.gov/StateFacts/NE.htm)
- Nebraska Frequently Asked Questions (http://nebraskaccess.ne.gov/NEfaq.asp)
- Nebraska State Publications Online (http://www.nlc.state.ne.us/docs/pilot/pilot.html)
- Nebraska city-data (http://www.city-data.com/city/Nebraska.html)
- nebraskastudies.org (http://www.nebraskastudies.org) – History of Nebraska from Nebraska Department of Education, Nebraska State Historical Society, and NET
- Nebraska State Databases (http://wikis.ala.org/godort/index.php/Nebraska) – Annotated list of searchable databases produced by Nebraska state agencies and compiled by the Government Documents Roundtable of the American Library Association.
- Nebraska (http://www.dmoz.org/Regional/North_America/United_States/Nebraska/) at the Open Directory Project

gag:Nebraska mrj:Небраска

Republican River

Republican River	
River	
 June 24, 1947 flood of the Republican River on the border of Jewell County, Kansas and Republic County, Kansas near Hardy, Nebraska and Webber, Kansas, just south of Nebraska NE-8 on Kansas 1 Rd/CR-1 bridge over the Republican River. The normal flood stage for the river is at the tree line in the foreground.	
Country	United States
States	Kansas, Nebraska
Tributaries	
- left	North Fork Republican River
- right	Arikaree River
Source	
- location	Dundy County, Nebraska
- coordinates	40°01′12″N 101°56′16″W [1]
Mouth	Kansas River
- location	Jefferson Township / Smoky Hill Township, Geary County, Kansas
- coordinates	39°3′36″N 96°48′5″W [1]
Length	453 mi (729 km) [2]

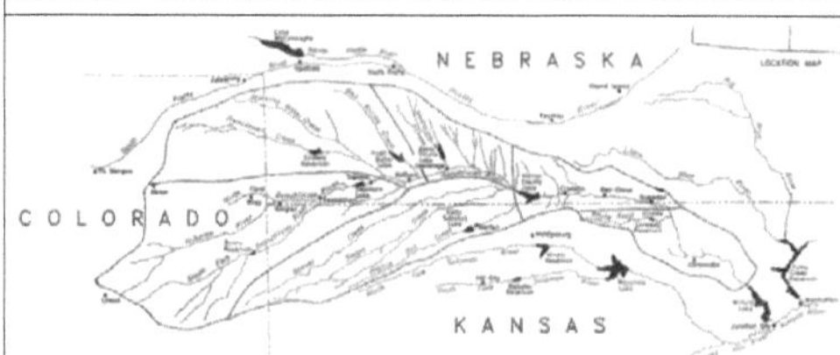

Map of the Republican River and its tributaries

The **Republican River** is a river in the central Great Plains of North America, flowing 453 miles (729 km)[2] through the U.S. states of Nebraska and Kansas.

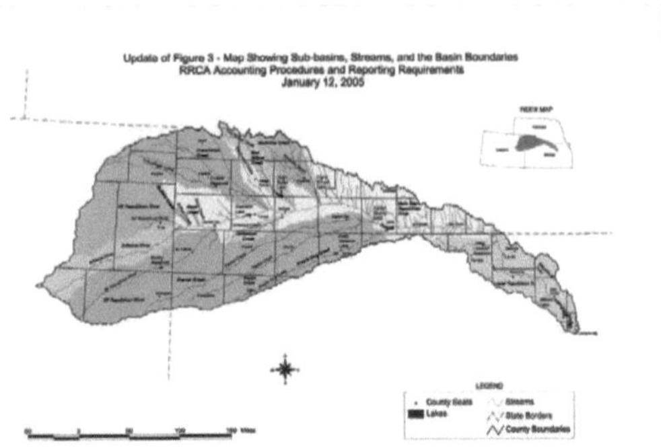

Republican River drainage basin

Geography

The Republican River is formed by the confluence of the North Fork Republican River and the Arikaree River just north of Haigler in Dundy County, Nebraska. It joins with the South Fork Republican River immediately southeast of Benkelman, Nebraska. All three tributaries originate in the High Plains of northeastern Colorado.[3] [4] [5] From the confluence, the river flows generally eastward along the southern border of Nebraska, passing through Swanson Reservoir and Harlan County Reservoir before curving southward into the Smoky Hills region of Kansas. The Republican River joins the Smoky Hill River at Junction City, Kansas to form the Kansas River.

Some cities along the river are McCook, Nebraska, Clay Center, Kansas, Concordia, Kansas and Junction City, Kansas. Near Concordia is the Republican River Pegram Truss, a bridge that goes over the Republican River that is listed on the National Register of Historic Places.

History

The river was named after a branch of Pawnee Indians known as "the Republicans" (Stewart 1967:223).[6] Allocation of the water from the Republican River is governed through an agreement called the Republican River Compact, involving the states of Nebraska, Kansas and Colorado.

Major flooding

July 1902

On July 9, 1902, the river flooded near Concordia, breaking a dam and re-routing the river by 1/4 of a mile.[7]

May/June 1935

The storm of May 31/June 1 (called "Nebraska's Deadliest Flood") dumped an average rainfall of nine inches on the river's watershed. This storm was also unique in that it moved in the same direction as the drainage basin. As a result, the Frenchman, Red Willow, Medicine, Deer, Muddy, and Turkey creeks all reached their flood peaks at the same time as the crest passed on the Republican River.

According to witness accounts, the roar of the water could be heard coming down the Republican Valley five miles away. Many survivors also reported that there were two crests - the water came up on May 28, then receded slightly, but the second crest on June 1 greatly exceeded the first. At one point, the water rose six feet in thirty minutes and was ten to fifteen feet higher than the previous record crest. Another account states that the Republican rose 10 feet in 12 minutes in McCook; naturally, anything in the path of that wall of water would be destroyed. Water was twenty feet deep in some places, and the discharge was an incredible 280,000 cubic feet/second - more than 320 times the normal flow today.

Estimates show 113 people killed. over 11,400 to 41,500 head of cattle were killed with one report stating that carcasses littered roads as to make them impassable. 341 miles (549 km) of highway and 307 bridges were destroyed with 74500 acres (301 km^2) of farmland were inundated.[8]

Popular culture

The river is mentioned as a landmark in the Chapter 30 of the book Around the World in 80 Days by Jules Verne.[9]

See also

- List of Kansas rivers
- List of Nebraska rivers

References

[1] "Republican River" (http://geonames.usgs.gov/pls/gnispublic/f?p=gnispq:3:::NO::P3_FID:485443). Geographic Names Information System, U.S. Geological Survey. . Retrieved 2011-02-01.

[2] U.S. Geological Survey. National Hydrography Dataset high-resolution flowline data. The National Map (http://viewer.nationalmap.gov/viewer/), accessed March 25, 2011

[3] "North Fork Republican River" (http://geonames.usgs.gov/pls/gnispublic/f?p=gnispq:3:::NO::P3_FID:183028). Geographic Names Information System, U.S. Geological Survey. . Retrieved 2011-02-01.

[4] "South Fork Republican River" (http://geonames.usgs.gov/pls/gnispublic/f?p=gnispq:3:::NO::P3_FID:833592). Geographic Names Information System, U.S. Geological Survey. . Retrieved 2011-02-01.

[5] "Arikaree River" (http://geonames.usgs.gov/pls/gnispublic/f?p=gnispq:3:::NO::P3_FID:835484). Geographic Names Information System, U.S. Geological Survey. . Retrieved 2011-02-01.

[6] Stewart, George R. (1967) *Names on the Land*. Boston: Houghton Mifflin Company.

[7] **A Proud Past... A Pictorial History of Concordia, Kansas**, by Bell, Rachel Lowrey (1998), Marceline, Missouri: D-Books Publishing.

[8] Nebraska State Historical Society (http://www.dnr.state.ne.us/floodplain/mitigation/1935flood.html) "Republican Flood of 1935-Nebraska's Deadliest Flood"

[9] **Around the World in 80 Days** by Jules Verne, first published 1873

External links

- State of Colorado's Republican River Webpage (http://water.state.co.us/wateradmin/RepublicanRiver.asp)
- State of Kansas' Republican River Webpage (http://www.ksda.gov/interstate_water_issues/content/142)
- State of Nebraska's Republican River Webpage (http://www.dnr.ne.gov/Republican/RepRiverImplementation.html)

Reservoir

A **reservoir** (pronounced either REZ-ur-vore or REZ-er-vwar, etymology: from French *réservoir* a "storehouse [1]) or an **artificial lake** is used to store water. Reservoirs may be created in river valleys by the construction of a dam or may be built by excavation in the ground or by conventional construction techniques such as brickwork or cast concrete.

The Jhonghua Dam on the Dahan River in Taoyuan County, Taiwan.

The term reservoir may also be used to describe underground reservoirs such as an oil or water well.

Types

Valley dammed reservoir

A dam constructed in a valley relies on the natural topography to provide most of the basin of the reservoir. Dams are typically located at a narrow part of a valley downstream of a natural basin. The valley sides act as natural walls with the dam located at the narrowest practical point to provide strength and the lowest practical cost of construction. In many reservoir construction projects people have to be moved and re-housed, historical artifacts moved or rare environments relocated. Examples include the temples of Abu Simbel[2] (which were moved before the construction of the Aswan Dam to create Lake Nasser from the Nile in Egypt) and the re-location of the village of Capel Celyn during the construction of Llyn Celyn.[3]

Lake Vyrnwy Reservoir. The dam spans the Vyrnwy Valley and was the first large stone dam built in the United Kingdom.

Construction of a reservoir in a valley will usually necessitate the diversion of the river during part of the build often through a temporary tunnel or by-pass channel.[4]

Stocks Reservoir in Lancashire, England.

In hilly regions reservoirs are often constructed by enlarging existing lakes. Sometimes in such reservoirs the new top water level exceeds the watershed height on one or more of the feeder streams such as at Llyn Clywedog in Mid Wales.[5] In such cases additional side dams are required to contain the reservoir.

Where the topography is poorly suited to a single large reservoir, a number of smaller reservoirs may be constructed in a chain such as in the River Taff valley where the three reservoirs Llwyn-on Reservoir, Cantref Reservoir and Beacons Reservoir form a chain up the valley.[6]

Bank-side reservoir

Where water is taken from a river of variable quality or quantity, bank-side reservoirs may be constructed to store the water pumped or siphoned from the river. Such reservoirs are usually built partly by excavation and partly by the construction of a complete encircling bund or embankment which may exceed 6 km in circumference.[7] Both the floor of the reservoir and the bund must have an impermeable lining or core, often made of puddled clay. The water stored in such reservoirs may have a residence time of several months during which time normal biological processes are able to substantially reduce many contaminants and almost eliminate any turbidity. The use of bank-side reservoirs also allows a water abstraction to be closed down for extended period at times when the river is unacceptably polluted or when flow conditions are very low due to drought. The London water supply system is one example of the use of bank-side storage for all the water taken from the River Thames and River Lee with many large reservoirs such as Queen Mary Reservoir visible along the approach to London Heathrow Airport.[7]

Service reservoir

Service reservoirs[8] store fully treated potable water close to the point of distribution. Many service reservoirs are constructed as water towers, often as elevated structures on concrete pillars where the landscape is relatively flat. Other service reservoirs are entirely underground, especially in more hilly or mountainous country. In the United Kingdom, Thames Water has many underground reservoirs built in the 1800s by the Victorians, most of which are lined with brick. A good example is the Honor Oak Reservoir, constructed between 1901 and 1909. When it was completed it was the largest brick built underground reservoir in the world[9] and is still one of the largest in Europe[10] . The reservoir now forms part of the Southern extension of the Thames Water Ring Main. The top of the reservoir has been grassed over and is now the Aquarias Golf Club[11] .

Service reservoirs perform several functions including ensuring sufficient head of water in the water distribution system and providing hydraulic capacitance in the system to even out peak demand from consumers enabling the treatment plant to run at optimum efficiency. Large service reservoirs can also be managed to so that energy costs in pumping are reduced by concentrating refilling activity at times of day when power costs are low.

History

Five thousand years ago, the craters of extinct volcanoes in Arabia were used as reservoirs by farmers for their irrigation water.[12]

Dry climate and water scarcity in India led to early development of water management techniques, including the building of a reservoir at Girnar in 3000 BC.[13] Artificial lakes dating to the 5th century BC have been found in ancient Greece.[14] An artificial lake in present-day Madhya Pradesh province of India, constructed in the 11th century, covered 650 square metres (7000 sq ft).[13]

In Sri Lanka large reservoirs have been created by ancient Sinhalese kings in order to save the water for irrigation. The famous Sri Lankan king Parākramabāhu I of Sri Lanka stated " do not let a drop of water seep into the ocean without benefiting mankind ". He created the reservoir named Parakrama Samudra(sea of King Parakrama),[15] which has astonished archaeologists.

Uses

Direct water supply

Gibson Reservoir, Montana

Many dammed river reservoirs and most bank-side reservoirs are used to provide the raw water feed to a water treatment plant which delivers drinking water through water mains. The reservoir does not simply hold water until it is needed; it can also be the first part of the water treatment process. The time the water is held for before it is released is known as the *retention time*. This is a design feature that allows particles and silts to settle out, as well as time for natural biological treatment using algae, bacteria and zooplankton that naturally live within the water. However natural limnological processes in temperate climate lakes produces temperature stratification in the water body which tends to partition some elements such as manganese and phosphorus into deep, cold anoxic water during the summer months. In the autumn and winter the lake becomes fully mixed again. During drought conditions, it is sometimes necessary to draw down the cold bottom water and the elevated levels of manganese in particular can cause problems in water treatment plants.[16]

Hydroelectricity

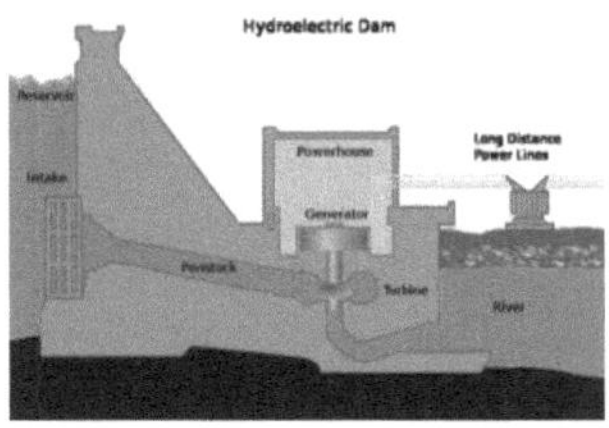

Hydroelectric dam in cross section.

A reservoir generating hydroelectricity includes turbines connected to the retained water body by large-diameter pipes. These generating sets may be at the base of the dam or some distance away. Some reservoirs generating hydroelectricity use pumped re-charge in which a high-level reservoir is filled with water using high-performance electric pumps at times when electricity demand is low and then uses this stored water to generate electricity by releasing the stored water into a low-level reservoir when electricity demand is high. Such systems are called pump-storage schemes.[17]

Controlling watercourses

Reservoirs can be used in a number of ways to control how water flows through downstream waterways.

Downstream water supply – water may be released from an upland reservoir so that it can be abstracted for drinking water lower down the system, sometimes hundred of miles further down downstream

Irrigation – water in an irrigation reservoir may be released into networks of canals for use in farmlands or secondary water systems. Irrigation may also be supported by reservoirs which maintain river flows allowing water to be abstracted for irrigation lower down the river.[18]

Flood control – also known as an "*attenuation*" or "*balancing*" reservoir, flood control reservoirs collect water at times of very high rainfall, then release it slowly over the course of the following weeks or months. Some of these reservoirs are constructed across the river line with the onward flow controlled by an orifice plate. When river flow exceeds the capacity of the orifice plate water builds behind the dam but as soon as the flow rate reduces the water behind the dam slowly releases until the reservoir is empty again. In some cases such reservoirs only function a few times in a decade and the land behind the reservoir may be developed as community or recreational land. A new generation of balancing dams are being developed to combat the climatic consequences of climate change. They are called "Flood Detention Reservoirs". Because these reservoirs will remain dry for long periods, there may be a risk of the clay core drying out reducing its structural stability. Recent developments include the use of composite core fill made from recycled materials as an alternative to clay.

Canals – Where a natural watercourse's water is not available to be diverted into a canal, a reservoir may be built to guarantee the water level in the canal; for example, where a canal climbs to cross a range of hills through locks.[19]

Recreation – water may be released from a reservoir to artificially create or supplement white-water conditions for kayaking and other white-water sports.[20] On salmonid rivers special releases (in Britain called *freshets*) are made to encourage natural migration behaviours in fish and to provide a variety of fishing conditions for anglers.

Recreational-only Kupferbach reservoir near Aachen/Germany.

Flow balancing

Reservoirs can be used to balance the flow in highly managed systems, taking in water during high flows and releasing it again during low flows. In order for this to work without pumping requires careful control of water levels using adjustable sluices. Accurate weather forecasts are essential so that dam operators can plan drawdowns prior to a high rainfall or snowmelt event. Dam operators blamed a faulty weather forecast on the 2010–2011 Queensland floods. Examples of highly managed Reservoirs are Burrendong Dam in Australia and Llyn Tegid in North Wales. Llyn Tegid is a natural lake whose level was raised by a low dam and into which the River Dee flows or discharges depending upon flow conditions at the time as part of the River Dee regulation system. This mode of operation is a form of hydraulic capacitance in the river system.

Recreation

The water bodies provided by many reservoirs often allow some recreational uses such as fishing, boating, and other activities. Special rules may apply for the safety of the public and to protect the quality of the water and the ecology of the surrounding area. Many reservoirs now support and encourage less informal and less structured recreation such as natural history, bird watching, landscape painting, walking and hiking and often provide information boards and interpretation material to encourage responsible use.

Operation

Water falling as rain upstream of the reservoir together with any groundwater emerging as springs is stored in the reservoir. Any excess water can be spilled via a specifically designed spillway. Stored water may be piped by gravity for use as drinking water, to generate hydro-electricity or to maintain river flows to support downstream uses. Occasionally reservoirs can be managed to retain high rain-fall events to prevent or reduce downstream flooding. Some reservoirs support several uses and the operating rules may be complex.

Most modern reservoirs have a specially designed draw-off tower that can discharge water from the reservoir at different levels both to access water as the reservoir draws down but also to allow water of a specific quality to be discharged into the downstream river as compensation water.

The operators of many upland or in-river reservoirs have obligations to release water into the downstream river to maintain river quality, support fisheries, maintain downstream industrial uses. maintain recreational use or for a range of other requirements. Such releases are known as *compensation water.*

Spillway of Llyn Brianne dam in Wales.

Terminology

The terminology for reservoirs varies from country to country. In most of the world reservoir areas are expressed in km^2 whilst in the USA acres are commonly used. For volume either m^3 or km^3 are widely used with acre feet used in the USA.

The capacity, volume or storage of a reservoir is usually divided into distinguishable areas. *Dead* or *inactive* storage refers to water in a reservoir that cannot be drained by gravity through a dam's outlet works, spillway or power plant intake and can only be pumped out. Dead storage allows sediments to settle which improves water quality and also creates hydraulic head along with an area for fish during low levels. *Active* or *live* storage is the portion of the reservoir that can be utilized for flood control, power production, navigation and downstream releases. In addition, a reservoir's *flood control capacity* is the amount of water it can regulate during flooding. The *surcharge capacity* is the capacity of the reservoir above the spillway crest that cannot be regulated.[21]

In the United States the water below the normal maximum level of a reservoir is called the *conservation pool.*[22]

In the UK *top water level* describes the reservoir full state whist *fully drawn down* describes the minimum retained volume.

Modelling reservoir management

There is a wide variety of software for modelling reservoirs, from the specialist Dam Safety Program Management Tools (DSPMT) to the relatively simple WAFLEX, to integrated models like the Water Evaluation And Planning system (WEAP) that place reservoir operations in the context of system-wide demands and supplies.

Safety

In many countries large reservoirs are closely regulated to try to prevent or minimise failures of containment.[23] [24]

Whilst much of the effort is directed at the dam and its associated structures as the weakest part of the overall structure, the aim of such controls is to prevent an uncontrolled release of water from the reservoir. Reservoir failures can generate huge increases in flow down a river valley with the potential to wash away towns and villages and cause considerable loss of life such as the devastation following the failure of containment at Llyn Eigiau which killed 17 people.[25] (see also List of dam failures)

A notable case of reservoirs being used as an instrument of War involved the British Royal Air Force Dambusters raid on Germany in World War II (codenamed "Operation Chastise" [26]), in which three German reservoir dams were selected to be breached in order to impact on German infrastructure and manufacturing and power capabilities deriving from the Ruhr and Eder rivers. The economic and social impact was derived from the enormous volumes of previously stored water that swept down the valleys wreaking destruction. This raid later became the basis for several films.

Environmental impact

Whole life environmental impact

All reservoirs will have a monetary cost/benefit assessment made before construction to see if the project is worth proceeding with.[27] However, such analysis can often omit the environmental impacts of dams and the reservoirs that they contain. Some impacts such as the greenhouse gas production associated with concrete manufacture are relatively easy to estimate. Other impact on the natural environment and social and cultural effects can be more difficult to assess and to weigh in the balance but identification and quantification of these issues are now commonly required in major construction projects in the developed world [28]

Climate change

Depending upon the circumstances, a reservoir built for hydro-electricity generation can either reduce or increase the net production of greenhouse gases. An **increase** can occur if plant material in the flooded areas decays in an anaerobic environment releasing (methane and carbon dioxide). This apparently counterintuitive position arises because much carbon is released as methane which is approximately 8 time more potent as a greenhouse gas than carbon dioxide [29]

A study for the National Institute for Research in the Amazon found that Hydroelectric reservoirs release a large pulse of carbon dioxide from above-water decay of trees left standing in the reservoirs, especially during the first decade after closing.[30] This elevates the global warming impact of the dams to levels much higher than would occur by generating the same power from fossil fuels.[30] According to the World Commission on Dams report (Dams And Development), when the reservoir is relatively large and no prior clearing of forest in the flooded area was undertaken, greenhouse gas emissions from the reservoir could be higher than those of a conventional oil-fired thermal generation plant.[31] For instance, In 1990, the impoundment behind the Balbina Dam in Brazil (inaugurated in 1987) had over 20 times the impact on global warming than would generating the same power from fossil fuels, due to the large area flooded per unit of electricity generated.[30]

A **decrease** can occur if the dam is used in place of traditional power generation, since electricity produced from hydroelectric generation does not give rise to any flue gas emissions from fossil fuel combustion (including sulfur dioxide, nitric oxide and carbon monoxide from coal). The Tucurui dam in Brazil (closed in 1984) had only 0.4 times the impact on global warming than would generating the same power from fossil fuels.[30]

Biology

Dams can produce a block for migrating fish, trapping them in one area, producing food and a habitat for various water-birds. They can also flood various ecosystems on land and may cause extinctions.

Human Impact

Dams can severely reduce the amount of water reaching countries downstream of them, causing water stress between the countries, e.g. the Sudan and Egypt, which damages farming businesses in the downstream countries, and reduces drinking water.

Farms and villages, e.g. Ashopton can be flooded by the creation of reservoirs, ruining many livelihoods. For this very reason, worldwide 80 million people (figure is as of 2009) have had to be forcibly relocated due to dam construction.

Limnology

The limnology of reservoirs has many similarities to that of lakes of equivalent size. There are however significant differences.[32] Many reservoirs experience considerable variations in level producing significant areas that are intermittently underwater or dried out. This greatly limits the productivity or the water margins and limits the number of species able to survive in these conditions.

Upland reservoirs tend to have a much shorter residence time than natural lakes and this can lead to more rapid cycling of nutrients through the water body so that they are more quickly lost to the system. This may be seen as a mismatch between water chemistry and water biology with a tendency for the biological component to be more oligotrophic than the chemistry would suggest.

Conversely, lowland reservoirs drawing water from nutrient rich rivers, may show exaggerated eutrophic characteristics because the residence time in the reservoir is much greater than in the river and the biological systems have a much greater opportunity to utilise the available nutrients.

Deep reservoirs with multiple level draw off towers can discharge deep cold water into the downstream river greatly reducing the size of any hypolimnion. This in turn can reduce the concentrations of phosphorus released during any annual mixing event and may therefore reduce productivity.

The Dams in front of reservoirs act as knickpoints-the energy of the water falling from them reduces and deposition is a result below the Dams.

Seismicity

The filling (impounding) of reservoirs has often been attributed to reservoir-triggered seismicity (RTS) as seismic events have occurred near large dams or within their reservoirs in the past. These events may have been triggered by the filling or operation of the reservoir and are on a small scale when compared to the amount of reservoirs worldwide. Of over 100 recorded events, early examples include the 60 m (197 ft) tall Marathon Dam in Greece (1929), the 221 m (725 ft) tall Hoover Dam in the U.S. (1935). Most events involve large dams and small amounts of seismicity. The only four recorded events above a 6.0-magnitude (M_w) are the 103 m (338 ft) tall Koyna Dam in India which registered a M_w of 6.3 along with the 120 m (394 ft) Kremasta Dam in Greece which registered a 6.3-M_w as well. Following those two, the next largest were the 122 m (400 ft) high Kariba Dam in Zambia at 6.25-M_w and the 105 m (344 ft) Xinfengjiang Dam in China at 6.1-M_w. Disputes occur over when RTS has occurred due to a lack of hydrogeological knowledge at the time of the event. It is accepted though that the infiltration of water into pores and the weight of the reservoir do contribute to RTS patterns. For RTS to occur, there must be a seismic structure near the dam or its reservoir and the seismic structure must be close to failure. Additionally, water must be able to infiltrate the deep rock stratum as the weight of a 100 m (328 ft) deep reservoir will have little impact when compared the deadweight of rock on a crustal stress field which may be located at a depth of 10 km (6 mi) or more.[33]

Micro climate

Reservoirs may change the local micro-climate increasing humidity and reducing extremes of temperature. Such effects are claimed by some South Australian winerys as increasing the quality of the wine production.

List of reservoirs

List of reservoirs by area

The following are the world's ten largest reservoirs by surface area:

1. Lake Volta (8482 km^2 or 3275 sq mi; Ghana) [34]
2. Smallwood Reservoir (6527 km^2 or 2520 sq mi; Canada)[35]
3. Kuybyshev Reservoir (6450 km^2 or 2490 sq mi; Russia)[36]
4. Lake Kariba (5580 km^2 or 2150 sq mi; Zimbabwe, Zambia)[37]
5. Bukhtarma Reservoir (5490 km^2 or 2120 sq mi; Kazakhstan)
6. Bratsk Reservoir (5426 km^2 or 2095 sq mi; Russia)[38]
7. Lake Nasser (5248 km^2 or 2026 sq mi; Egypt, Sudan) [39]
8. Rybinsk Reservoir (4580 km^2 or 1770 sq mi; Russia)
9. Caniapiscau Reservoir (4318 km^2 or 1667 sq mi; Canada)[40]
10. Lake Guri (4250 km^2 or 1640 sq mi; Venezuela)

Lake Volta from space (April 1993).

List of reservoirs by volume

1. Lake Kariba (180 km^3 or 43 cu mi; Zimbabwe, Zambia)
2. Bratsk Reservoir (169 km^3 or 41 cu mi; Russia)
3. Lake Nasser (157 km^3 or 38 cu mi; Egypt, Sudan)
4. Lake Volta (148 km^3 or 36 cu mi; Ghana)
5. Manicouagan Reservoir (142 km^3 or 34 cu mi; Canada)[41]
6. Lake Guri (135 km^3 or 32 cu mi; Venezuela)
7. Williston Lake (74 km^3 or 18 cu mi; Canada)[42]
8. Krasnoyarsk Reservoir (73 km^3 or 18 cu mi; Russia)
9. Zeya Reservoir (68 km^3 or 16 cu mi; Russia)

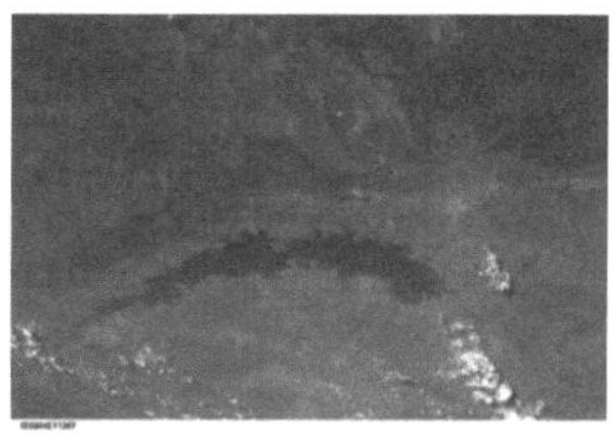

Lake Kariba from space.

See also

- Ab Anbar
- Colourful lakelets (in Poland)
- Drainage basin
- Drought
- Hydroelectricity
- Dam failure
- Mill pond
- Multipurpose reservoir
- Spillway
- Coastal sediment supply

References

[1] Online Etymology Dictionary – Reservoir (http://www.etymonline.com/index.php?term=reservoir)
[2] Nubian Monuments from Abu Simbel to Philae – UNESCO World Heritage Centre (http://whc.unesco.org/en/list/88)
[3] *Capel Celyn, Ten Years of Destruction: 1955–1965*, Thomas E., Cyhoeddiadau Barddas & Gwynedd Council, 2007, ISBN 978 1 900437 92 9
[4] Construction of Hoover Dam: a historic account prepared in cooperation with the Department of the Interior. KC Publications. 1976. ISBN 0-916122-51-4.
[5] Llyn Clywedog – Llanidloes mid-Wales (http://www.llanidloes.com/clywedog/index.html)
[6] Reservoirs of Fforest Fawr Geopark (http://www.geoparcyfforestfawr.org.uk/understanding/archaeology-industrial-heritage/reservoir-of-fforest-fawr-geopark)
[7] Queen Mary and King George V emergency draw down schemes (http://www.icevirtuallibrary.com/content/article/10.1680/dare.2009.19.2.79;jsessionid=3295v7olnvqlv.z-telford-01)
[8] Open University – Service Reservoirs (http://openlearn.open.ac.uk/mod/resource/view.php?id=185924)
[9] "Honor Oak Reservoir" (http://www.lewisham.gov.uk/SiteCollectionDocuments/ForestHillAndHonorOakSecretsLeaflet.pdf). London Borough of Lewisham. . Retrieved 2011-09-01.
[10] "Honor Oak Reservoir" (http://www.projectmanagement.mottmac.com/projects/?mode=type&id=130093). Mott MacDonald. . Retrieved 2011-09-01.
[11] Golf Club website (http://www.aquariusgolfclub.co.uk/pages.php/index.htmlAquarius)
[12] Smith, S. et al. (2006) *Water: the vital resource*, 2nd edition, Milton Keynes, The Open University
[13] edited by John C. Rodda, Lucio Ubertini. (2004). Rodda, John; Ubertini, Lucio. eds. *The Basis of Civilization – Water Science?*. International Association of Hydrological Science. ISBN 1-901502-57-0. OCLC 224463869
[14] Wilson & Wilson (2005). *Encyclopedia of Ancient Greece*. Routledge. ISBN 0415973341. pp. 8
[15] – International Lake Environment Committee – Parakrama Samudra (http://www.ilec.or.jp/database/asi/asi-45.html)
[16] Water problems – Manganese (http://www.freedrinkingwater.com/water_quality/chemical/water-problems-manganese.htm)
[17] How pump storage works (http://www.fhc.co.uk/pumped_storage.htm)
[18] Thinking about an irrigation reservoir? (http://www.ukia.org/eabooklets/EA Reservoir booklet_final.pdf)
[19] Huddersfield narrow canal reservoirs (http://www.huddersfield1.co.uk/huddersfield/narrowcanal/huddscanalres.htm)
[20] Water Release information for The River Tryweryn at the National Whitewater centre (http://www.ukrafting.co.uk/waterinfo.htm)
[21] Vojtěch Broža, Ladislav Votruba, (1989). *Water management in reservoirs* (http://books.google.com/?id=j8dIlPJITH0C&pg=PA187&lpg=PA187&dq=active+storage+reservoir#v=onepage&q=active storage reservoir&f=false). Elsevier Publishing Company. p. 187. ISBN 0-444-98933. .
[22] Lower Colorado River Authority – Water Glossary (http://www.lcra.org/water/conditions/glossary.html)
[23] North Carolina Dam safety law (http://www.dlr.enr.state.nc.us/pages/damsafetylaw1967.html)
[24] Reservoirs Act 1975 The Reservoirs Act 1975 (UK) (http://www.opsi.gov.uk/RevisedStatutes/Acts/ukpga/1975/cukpga_19750023_en_1)
[25] Snowdonia – Llyn Eigau (http://www.snowdoniaguide.com/llyn_eigiau.html)
[26] Commonwealth War Graves Commission – Operation Chastise (http://www.cwgc.org/admin/files/The Dams Raid.pdf)
[27] CIWEM – Reservoirs:Global Issues (http://www.ciwem.org/policy/policies/reservoirs.asp)
[28] Proposed reservoir – Environmental Impact Assessment (EIA) Scoping Report (http://www.whitehorsedc.gov.uk/news_views/topical_issues/xx_detailpage-4402.asp)
[29] Houghton, John (4 May 2005). "Global warming" (http://stacks.iop.org/RoPP/68/1343). *Reports on Progress in Physics* (Institute of Physics) **68** (6): 1362. doi:10.1088/0034-4885/68/6/R02. .
[30] Fearnside, P.M. 1995. hydroelectric dams in the Brazilian Amazon as sources of 'greenhouse' gases. Environmental Conservation 22(1): 7–19.]
[31] Hydroelectric power's dirty secret revealed – earth – 24 February 2005 – New Scientist (http://www.newscientist.com/article.ns?id=dn7046)
[32] Ecology of Reservoirs and Lakes (http://www.forestencyclopedia.net/p/p1483)
[33] "The relationship between large reservoirs and seismicity 08 February 2010" (http://www.waterpowermagazine.com/story.asp?storyCode=2055399). International Water Power & Dam Construction. 20 February 2010. . Retrieved 12 March 2011.
[34] International Lake Environment Committee – Volta Lake (http://www.ilec.or.jp/database/afr/afr-16.html)
[35] The Canadian Encyclopaedia – Smallwood Reservoir (http://www.thecanadianencyclopedia.com/index.cfm?PgNm=TCE&Params=A1ARTA0007465)
[36] International Lake Environment Committee – Reservoir Kuybyshev (http://www.ilec.or.jp/database/eur/deur54.html)
[37] International Lake Environment Committee – Lake Kariba (http://www.ilec.or.jp/database/afr/dafr04.html)
[38] International Lake Environment Committee – Bratskoye Reservoir (http://www.ilec.or.jp/database/asi/dasi61.html)
[39] International Lake Environment Committee – Aswam high dam reservoir (http://www.ilec.or.jp/database/afr/dafr19.html)
[40] International Lake Environment Committee – Caniapiscau Reservoir (http://www.ilec.or.jp/database/nam/dnam35.html)
[41] International Lake Environment Committee – Manicouagan Reservoir (http://www.ilec.or.jp/database/nam/dnam26.html)
[42] International Lake Environment Committee – Williston Lake (http://www.ilec.or.jp/database/nam/dnam29.html)

External links

- Department of Water Resources. "Reservoir Information" (http://cdec.water.ca.gov/misc/resinfo.html). *California Data Exchange Center*. State of California.

ltg:Iudiųture

Phillips County, Kansas

Phillips County, Kansas

Phillips County Courthouse

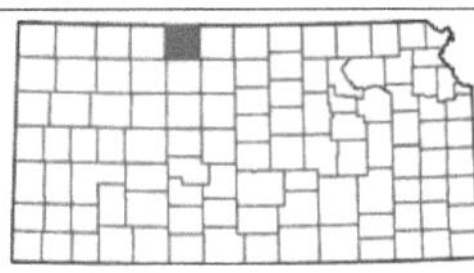

Location in the state of Kansas

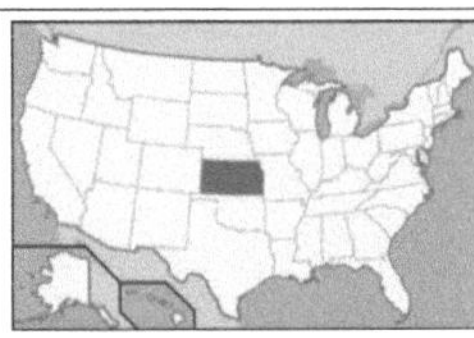

Kansas's location in the U.S.

Founded	February 26, 1867
Seat	Phillipsburg
Area **- Total** **- Land** **- Water**	894.81 sq mi (2318 km²) 886.23 sq mi (2295 km²) 8.58 sq mi (22 km²), 0.96%
Population **- (2010)** **- Density**	5642 6.2/sq mi (2.4/km²)
Time zone	Central: UTC-6/-5
Website	www.PhillipsCounty.org [1]

Phillips County (standard abbreviation: **PL**) is a county located in the U.S. state of Kansas. As of the 2010 census, the county population was 5,642.[2] The largest city and county seat is Phillipsburg.

Law and government

Following amendment to the Kansas Constitution in 1986, the county remained a prohibition, or "dry", county until 1996, when voters approved the sale of alcoholic liquor by the individual drink with a 30% food sales requirement.[3]

Geography

According to the 2000 census, the county has a total area of 894.81 square miles (2317.5 km^2), of which 886.23 square miles (2295.3 km^2) (or 99.04%) is land and 8.58 square miles (22.2 km^2) (or 0.96%) is water.[4]

Adjacent counties

- Harlan County, Nebraska (north)
- Franklin County, Nebraska (northeast)
- Smith County (east)
- Rooks County (south)
- Graham County (southwest)
- Norton County (west)

National protected area

- Kirwin National Wildlife Refuge

Demographics

Historical populations		
Census	Pop.	%±
1900	14442	—
1910	14150	−2.0%
1920	12505	−11.6%
1930	12159	−2.8%
1940	10435	−14.2%
1950	9273	−11.1%
1960	8709	−6.1%
1970	7888	−9.4%
1980	7406	−6.1%
1990	6590	−11.0%
2000	6001	−8.9%
2010	5642	−6.0%
U.S. Decennial Census [5]		

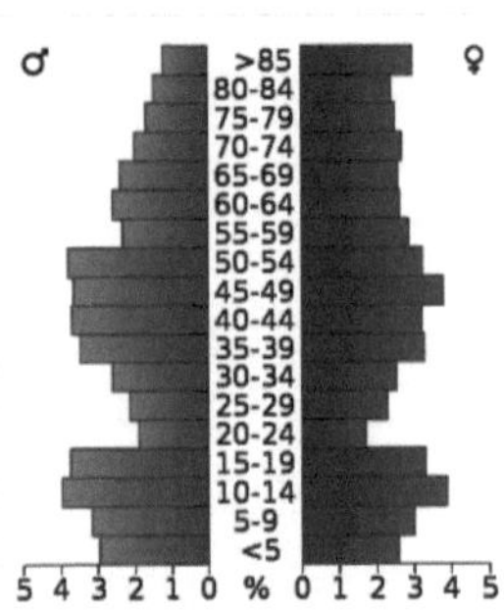

Age pyramid

As of the census[6] of 2000, there were 6,001 people, 2,496 households, and 1,722 families residing in the county. The population density was 7 people per square mile (3/km²). There were 3,088 housing units at an average density of 4 per square mile (1/km²). The racial makeup of the county was 98.25% White, 0.25% Black or African American, 0.30% Native American, 0.45% Asian, 0.03% from other races, and 0.72% from two or more races. 0.67% of the population were Hispanic or Latino of any race.

There were 2,496 households out of which 28.30% had children under the age of 18 living with them, 61.50% were married couples living together, 5.50% had a female householder with no husband present, and 31.00% were non-families. 28.60% of all households were made up of individuals and 15.80% had someone living alone who was 65 years of age or older. The average household size was 2.35 and the average family size was 2.89.

In the county the population was spread out with 24.50% under the age of 18, 5.70% from 18 to 24, 23.20% from 25 to 44, 24.80% from 45 to 64, and 21.80% who were 65 years of age or older. The median age was 42 years. For every 100 females there were 94.80 males. For every 100 females age 18 and over, there were 91.20 males.

The median income for a household in the county was $35,013, and the median income for a family was $41,638. Males had a median income of $29,609 versus $17,827 for females. The per capita income for the county was $17,121. About 7.20% of families and 10.00% of the population were below the poverty line, including 11.60% of those under age 18 and 8.80% of those age 65 or over.

Cities and towns

Incorporated cities

Name and population (2004 estimate):

- Phillipsburg, 2,473 (county seat)
- Logan, 555
- Agra, 283
- Kirwin, 217
- Long Island, 146
- Prairie View, 133
- Glade, 108
- Speed, 42

Unincorporated places

- Stuttgart
- Woodruff

Townships

Phillips County is divided into twenty-five townships. The city of Phillipsburg is considered *governmentally independent* and is excluded from the census figures for the townships. In the following table, the population center is the largest city (or cities) included in that township's population total, if it is of a significant size.

Sources: 2000 U.S. Gazetteer [7] from the U.S. Census Bureau.

Township	FIPS	Population center	Population	Population density /km² (/sq mi)	Land area km² (sq mi)	Water area km² (sq mi)	Water %	Geographic coordinates
Arcade	02150		96	1 (3)	92 (36)	0 (0)	0.03%	39°46′30″N 99°13′40″W
Beaver	05100		54	1 (2)	93 (36)	0 (0)	0.05%	39°47′22″N 99°33′43″W
Belmont	05700		113	1 (3)	93 (36)	0 (0)	0 %	39°41′26″N 99°27′13″W
Bow Creek	08075		44	0 (1)	92 (35)	1 (1)	1.49%	39°36′15″N 99°13′31″W
Crystal	16550		59	1 (2)	92 (36)	0 (0)	0.19%	39°51′55″N 99°6′35″W
Dayton	17100		53	1 (1)	92 (36)	0 (0)	0.10%	39°53′27″N 99°27′15″W
Deer Creek	17200		67	1 (2)	88 (34)	5 (2)	4.87%	39°41′2″N 99°13′54″W
Freedom	24650		85	1 (2)	92 (36)	0 (0)	0.10%	39°52′18″N 99°21′40″W
Glenwood	26650		42	0 (1)	92 (36)	0 (0)	0.21%	39°57′38″N 99°14′25″W
Granite	27450		21	0 (1)	92 (36)	0 (0)	0.14%	39°58′6″N 99°28′11″W
Greenwood	28775		45	0 (1)	92 (36)	0 (0)	0.05%	39°51′51″N 99°14′24″W
Kirwin	37300		307	4 (9)	87 (33)	6 (2)	6.89%	39°40′51″N 99°7′24″W
Logan	42125		666	7 (19)	93 (36)	0 (0)	0.03%	39°40′3″N 99°34′11″W
Long Island	42625		251	3 (7)	91 (35)	0 (0)	0.12%	39°56′35″N 99°33′16″W
Mound	48725		129	1 (4)	93 (36)	0 (0)	0 %	39°47′13″N 99°27′7″W
Phillipsburg	55700		257	3 (7)	89 (34)	0 (0)	0.01%	39°46′15″N 99°20′28″W
Plainview	56125		23	0 (1)	93 (36)	0 (0)	0.01%	39°35′21″N 99°27′1″W
Plum	56775		447	5 (12)	93 (36)	0 (0)	0.27%	39°46′3″N 99°7′6″W
Prairie View	57550		220	2 (6)	92 (36)	0 (0)	0.13%	39°51′0″N 99°34′38″W
Rushville	61800		23	0 (1)	93 (36)	0 (0)	0 %	39°36′40″N 99°21′35″W
Solomon	66350		226	2 (6)	92 (36)	0 (0)	0.05%	39°41′27″N 99°20′12″W

Sumner	69175		43	0 (1)	92 (35)	0 (0)	0.18%	39°57′6″N 99°6′50″W
Towanda	71175		26	0 (1)	93 (36)	0 (0)	0.20%	39°36′13″N 99°33′43″W
Valley	73025		22	0 (1)	85 (33)	8 (3)	8.70%	39°36′58″N 99°8′3″W
Walnut	75075		14	0 (0)	93 (36)	0 (0)	0.09%	39°57′33″N 99°20′47″W

Education

Unified school districts

- Thunder Ridge USD 110 [8] (Consolidation of Agra-Eastern Heights and Kensington-West Smith County)
- Phillipsburg USD 325 [9]
- Logan USD 326 [10]

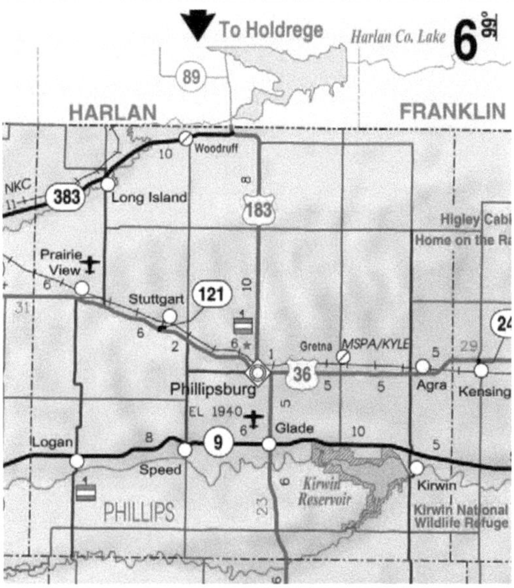

2005 KDOT Map of Phillips County (map legend)

See also

- National Register of Historic Places listings in Phillips County, Kansas

Information on this and other counties in Kansas

- List of counties in Kansas
- National Register of Historic Places listings in Kansas
- Kansas locations by per capita income

Other information for Kansas

- List of cities in Kansas
- List of unified school districts in Kansas
- List of colleges and universities in Kansas

References

[1] http://www.PhillipsCounty.org

[2] "2010 County Population and Housing Occupancy Status" (http://factfinder2.census.gov/faces/tableservices/jsf/pages/productview.xhtml?pid=DEC_10_PL_GCTPL2.ST05&prodType=table). U.S. Census Bureau. . Retrieved April 5, 2011.

[3] "Map of Wet and Dry Counties" (http://www.ksrevenue.org/abcwetdrymap.htm). Alcoholic Beverage Control, Kansas Department of Revenue. November 2006. . Retrieved 2007-12-26.

[4] "Census 2000 U.S. Gazetteer Files: Counties" (http://www.census.gov/tiger/tms/gazetteer/county2k.txt). United States Census. . Retrieved 2011-02-13.

[5] http://www.census.gov/prod/www/abs/decennial/

[6] "American FactFinder" (http://factfinder.census.gov). United States Census Bureau. . Retrieved 2008-01-31.

[7] http://www.census.gov/geo/www/gazetteer/places2k.html

[8] http://www.usd110.net/

[9] http://usd325.com/

[10] http://www.usd326.k12.ks.us/

Further reading

- *History of the State of Kansas*; William G. Cutler; A.T. Andreas Publisher; 1883. (Online HTML eBook) (http://www.kancoll.org/books/cutler/)
- *Kansas : A Cyclopedia of State History, Embracing Events, Institutions, Industries, Counties, Cities, Towns, Prominent Persons, Etc*; 3 Volumes; Frank W. Blackmar; Standard Publishing Co; 944 / 955 / 824 pages; 1912. (Volume1 - Download 54MB PDF eBook) (http://www.archive.org/download/kansascyc01blac/kansascyc01blac.pdf), (Volume2 - Download 53MB PDF eBook) (http://www.archive.org/download/kansascycloped02blac/kansascycloped02blac.pdf), (Volume3 - Download 33MB PDF eBook) (http://www.archive.org/download/kansascyclopedia03blac/kansascyclopedia03blac.pdf)

External links

Official websites

- Phillips County (http://www.PhillipsCounty.org/)

General county information

- Blue Skyways (http://skyways.lib.ks.us/counties/PL/)

County Level Data

- Kansas Statistical Abstract (http://www.ku.edu/pri/ksdata/county.shtml)

Maps

- Phillips County Map (http://www.ksdot.org/burtransplan/maps/county-pdf/phillips.PDF), KDOT
- Kansas Highway Map (http://www.ksdot.org/burtransplan/maps/state-pdf/2011-12Mapside.pdf), KDOT
- Kansas Railroad Map (http://www.ksdot.org/BurTransPlan/maps/SpecialInterestStateMaps/Railroad2011.pdf), KDOT
- Kansas School District Boundary Map (http://www3.ksde.org/eddir/usd_map.pdf), KSDE

Colorado

State of Colorado	
 Flag  Seal	
Nickname(s): *The Centennial State*	
Motto(s): *Nil sine numine (Nothing without providence)*	

Capital (and largest city)	Denver
Largest metro area	Denver-Aurora-Broomfield MSA
Area	Ranked 8th in the U.S.
- Total	104,094 sq mi (269,837 km^2)
- Width	380 miles (612 km)
- Length	280 miles (451 km)
- % water	0.36%
- Latitude	37°N to 41°N
- Longitude	102°03'W to 109°03'W
Population	Ranked 22nd in the U.S.
- Total	(2010) 5,029,196
- Density	48.31/sq mi (18.64/km^2) Ranked 37th in the U.S.
- Median income	$56,993 (13th)
Elevation	
- Highest point	Mount Elbert[1] [2] 14,440 ft (4401.2 m)

- Mean	6,800 ft (2073 m)
- Lowest point	Arikaree River[1] 3,317 ft (1011 m)
Admission to Union	August 1, 1876 (38th State)
Governor	John Hickenlooper (D)
Lieutenant Governor	Joseph A. Garcia (D)
Legislature	General Assembly
- Upper house	Senate
- Lower house	House of Representatives
U.S. Senators	2 – Mark Udall (D) 3 – Michael Bennet (D)
U.S. House delegation	4 Republicans and 3 Democrats (list)
Time zone	MST=UTC-07, MDT=UTC-06
Abbreviations	CO Colo. US-CO
Website	[www.colorado.gov www.colorado.gov]

Colorado (pronounced i/kɒləˈrædoʊ/ or, i/kɒləˈrɑːdoʊ/)[3] is a U.S. state that encompasses much of the Rocky Mountains as well as the northeastern portion of the Colorado Plateau and the western edge of the Great Plains. Colorado is part of the Western United States and the Mountain States.

The state was named for the Colorado River, which early Spanish explorers named the *Río Colorado* for the red colored (Spanish: *colorado*) silt the river carried from the mountains. On August 1, 1876, U.S. President Ulysses S. Grant signed a proclamation admitting Colorado as the 38th state. Colorado is nicknamed the "Centennial State" because it was admitted to the Union as the 38th state in 1876, the centennial year of the United States Declaration of Independence.

Colorado is bordered by the northwest state of Wyoming to the north, the midwest states of Nebraska and Kansas to the northeast and east, on the south by New Mexico and a small portion of the southern state of Oklahoma, and on the west by Utah. The four states of Colorado, New Mexico, Arizona, and Utah meet at one common point known as the Four Corners, which is known as the heart of the American Southwest. Colorado is one of only three U.S. states with no natural borders, the others being neighboring Wyoming and Utah.

Colorado is noted for its vivid landscape of mountains, forests, high plains, mesas, canyons, plateaus, rivers, and desert lands. The 2010 United States Census tallied the state population at 5,029,196 as of April 1, 2010, an increase of 16.92% since the 2000 United States Census. Denver is the capital and the most populous city of Colorado. Residents of the state are properly known as "Coloradans", although the archaic term "Coloradoan" is still used.[4] [5]

Geography

The borders of Colorado were originally defined to be lines of latitude and longitude, making its shape a latitude-longitude* quadrangle which stretches from 37°N to 41°N latitude and from 102°03'W to 109°03'W longitude (25°W to 32°W from the Washington Meridian).[6] Colorado, Wyoming and Utah are the only states which have boundaries defined solely by lines of latitude and longitude. When placing the border markers for the Territory of Colorado, minor surveying errors resulted in several nearly imperceptible kinks, most notably along the border [7] with the Territory of Utah. Once agreed upon by the federal, state, and territorial governments, those surveyors' benchmarks, became the legal boundaries for the Colorado Territory, kinks and all.[8]

Tenmile Range near Leadville, Colorado

The summit of Mount Elbert at 14440 feet (4401 m) in elevation in Lake County is the highest point of Colorado and the Rocky Mountains.[2] Colorado is the only U.S. state that lies entirely above 1000 meters (3281 ft) elevation. The point where the Arikaree River flows out of Yuma County, Colorado, and into Cheyenne County, Kansas, is the lowest point in Colorado at 3317 feet (1011 m) elevation. This point holds the distinction of being the highest low elevation point of any state.[1] [9]

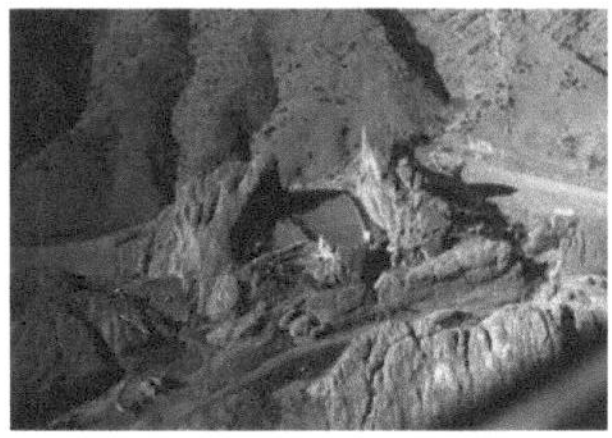
Red Rocks Amphitheatre, west of Denver, where the Plains meet the Rockies.

A little over one fourth of the area of Colorado is flat and rolling land. East of the Rocky Mountains are the Colorado Eastern Plains of the High Plains, the section of the Great Plains within Colorado at elevations ranging from roughly 3350 to 7360 feet (1020 to 2240 m). The midwest plains states of Kansas and Nebraska border Colorado to the east and northeast. The Colorado plains are usually thought of as prairies, but actually they have many patches of deciduous forests, buttes, and canyons, much like the high plains in New Mexico as well. Eastern Colorado is presently mainly covered in farmland, along with small farming villages and towns. Precipitation is fair, averaging from 15 to 25 inches (380 to 630 mm) annually.[10] The summers in the plains are normally hot and dry, while the winters are often quite cold, snowy, and icy. Corn, wheat, hay, soybeans, and oats are all typical crops, and most of the villages and towns in this region boast both a water tower and a grain elevator. As well as the farming of crops, Eastern Colorado has a good deal of livestock raising, such as at cattle ranches and hog farms and irrigation water is available from the South Platte, the Arkansas River, and a few other streams, and also from subterranean sources, including artesian wells. However, heavy use of ground water from wells for irrigation has caused underground water reserves to decline.

The high desert terrain of the San Luis Valley in southern Colorado.

Most of Colorado's population resides along the eastern edge of the Rocky Mountains in the Front Range Urban Corridor between Cheyenne, Wyoming, and Pueblo, Colorado. This region is partially protected from prevailing storms that blow in from the Pacific Ocean region by the high Rockies in the middle of Colorado. The only other significant population centers are at Grand Junction and Durango in far western Colorado.

The Continental Divide dips down to 11990 feet (3655 m) at Loveland Pass.

To the west of Great Plains of Colorado rises the eastern slope of the Rocky Mountains. Notable peaks of the Rocky Mountains include Longs Peak, Mount Evans, Pikes Peak, and the Spanish Peaks near Walsenburg, in southern Colorado. This area drains to the east and the southeast, ultimately either via the Mississippi River or the Rio Grande into the Gulf of Mexico.

The Continental Divide extends along the crest of the Rocky Mountains. The area of Colorado to the west of the Continental Divide is called the Western Slope of Colorado. Drainage water west of the Continental Divide flows to the southwest via the Green River and the Colorado River into the Gulf of California.

Within the interior of the Rocky Mountains are several large so-called "parks" or high broad basins. In the north, on the east side of the Continental Divide is the North Park of Colorado. The North Park is drained by the North Platte River, which flows north into Wyoming and Nebraska. Just to the south of North Park, but on the western side of the Continental Divide, is the Middle Park of Colorado, which is drained by the Colorado River. The South Park of Colorado is the region of the headwaters of the South Platte River.

In southmost Colorado is the large San Luis Valley, where the headwaters of the Rio Grande are located. The valley sits between the Sangre De Cristo Mountains and San Juan Mountains, and consists of large desert lands that eventually run into the mountains. The Rio Grande drains due south into New Mexico, Mexico, and Texas. Across the Sangre de Cristo Range to the east of the San Luis Valley lies the Wet Mountain Valley. These basins, particularly the San Luis Valley, lie along the Rio Grande Rift, a major geological formation of the Rocky Mountains, and its branches.

The Rocky Mountains within Colorado contain about 54 peaks that are 14000 feet (4267 m) or higher in elevation above sea level, known as fourteeners.[11] These mountains are largely covered with trees such as conifers and aspens up to the tree line, at an elevation of about 12140 feet (3700 m) in southern Colorado to about 10500 feet (3200 m) in northern Colorado. Above this only alpine vegetation grows. Only small parts of the Colorado Rockies are snow-covered year round. Much of the alpine snow melts by mid-August with the exception of a few snowcapped peaks and a few small glaciers. The Colorado Mineral Belt, stretching from the San Juan Mountains in the southwest to Boulder and Central City on the front range, contains most of the historic gold- and silver-mining districts of Colorado.

The 30 highest major summits of the Rocky Mountains of North America all lie within the state. Colorado is home to four national parks, six national monuments, two national recreation areas, two national historic sites, three national historic trails, a national scenic trail, 11 national forests, two national grasslands, 41 national wilderness areas, two national conservation areas, eight national wildlife refuges, 44 state parks, a state forest, 323 state wildlife areas, and numerous other scenic, historic, and recreational attractions.

Mount Elbert, the tallest peak in the Rocky Mountains.

The Western Slope of Colorado is drained by the Colorado River and its tributaries (primarily the Green River and the San Juan River), or by evaporation in its arid areas. Prominent in the southwestern area of the Western Slope are the high San Juan Mountains, a rugged mountain range, and to the west of the San Juan Mountains, the Colorado Plateau, a high arid region that borders Southern Utah. The city of Grand Junction, Colorado, is the largest city on the Western Slope, Grand Junction and Durango are the only major centers of radio and television broadcasting, newspapers, and higher education on the Western Slope. The Mesa State College in

Grand Junction, Western State College in Gunnison, and Fort Lewis College in Durango are the only four-year colleges in Colorado west of the Continental Divide.

August, 1964. Continental Divide at the top of Loveland Pass

Grand Junction is located along Interstate 70, the only major highway of Western Colorado. Grand Junction is also along the major railroad of the Western Slope, the Union Pacific, which also provides the tracks for Amtrak's California Zephyr passenger train, which crosses the Rocky Mountains between Denver and Grand Junction via a route on which there are no continuous highways.

To the southeast of Grand Junction is the Grand Mesa, said to be the world's largest flat-topped mountain. Other towns of the Western Slope include Glenwood Springs with its resort hot springs, and the ski resorts of Aspen, Breckenridge, Vail, Crested Butte, Steamboat Springs, and Telluride.

Grand Mesa, the largest mesa in the world.

The northwestern corner of Colorado is a sparsely-populated region, and it contains part of the noted Dinosaur National Monument, which is not only a paleontological area, but is also a scenic area of high, rocky hills, canyons, and streambeads. Here, the Green River briefly crosses over into Colorado.

From west to east, the land of Colorado consists of desert lands and desert plateaus, then alpine mountains with National Forests, then some scattered desert land in the southern mountain areas in the state, and then the relatively-flat grasslands and scattered forests of the Great Plains. The famous Pikes Peak is located just west of Colorado Springs. Its isolated peak is visible from nearly the Kansas border on clear days, and also far to the north and the south.[12] The desert lands in Colorado are located in and around areas such as, the Royal Gorge, Pueblo, Canon City, Florence, Great Sand Dunes National Park and Preserve, San Luis Valley, Cortez, Canyon of the Ancients National Monument, Hovenweep National Monument, Ute Mountain, Delta, Grand Junction, Colorado National Monument, Roan Plateau, and other areas surrounding the Uncompahgre Plateau and Uncompahgre National Forest.

Colorado is one of four states in the United States that share a common geographic point the Four Corners, together with Arizona, New Mexico, and Utah. At this intersection, it is possible to stand in four states at once.

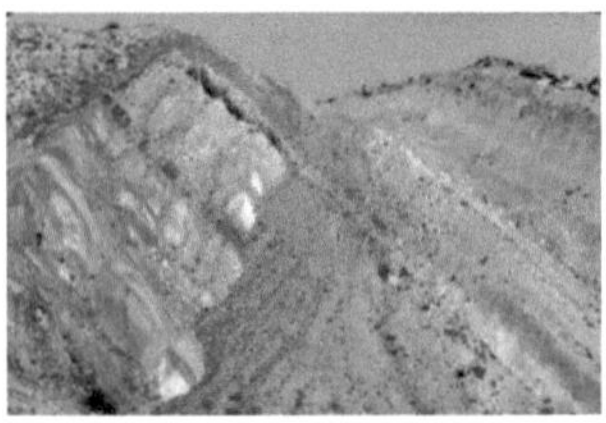

Dinosaur National Monument.

Climate

Spring melt at Bear Lake in Rocky Mountain National Park.

The climate of Colorado is quite complex compared to most of the United States. Unlike in other states, southern Colorado is not necessarily warmer than northern Colorado. Most of Colorado is made up of mountains, foothills, high plains, and desert lands. Mountains and surrounding valleys greatly affect local climate. As a general rule, with an increase in elevation comes a decrease in temperature and an increase in precipitation. Northeast, east, and southeast Colorado are mostly the high plains, while Northern Colorado is a mix of high plains, foothills, and mountains. Northwest and west Colorado are predominantly mountainous, with some desert lands mixed in. Southwest and southern Colorado are a complex mixture of desert and mountain areas.

Eastern Plains

Center pivot irrigation of wheat growing in Yuma County.

The climate of the Eastern Plains is semi-arid (Köppen climate classification *BSk*) with low humidity and moderate precipitation, usually from 15 to 25 inches (380 to 630 mm) annually. The area is known for its abundant sunshine and cool clear nights, which give this area a great average diurnal temperature range. In summer, this area can have many days above 95 °F (35 °C) and sometimes 100 °F (38 °C),[13] although 105 °F (41 °C) is the maximum in the front range cities above 5000 ft (1500 m). In the plains, the winter extremes can be from 0 °F (-18 °C) to −10 °F (−23 °C) and −15 °F (−26 °C). About 75% of the precipitation falls within the growing season, from April to September, but this area is very prone to droughts. Most of the precipitation comes from thunderstorms, which are often severe, and from major snowstorms that occur most often in the early spring, late autumn, and sometimes winter. Otherwise, winters tend to be mostly dry and cold. In much of the region, March and April are the snowiest months. April and May are normally the rainiest months, while April is the wettest month overall. The Front Range cities closer to the mountains tend to be warmer in the winter due to chinook winds which warm the area, sometimes bringing temperatures of 60 °F (16 °C) or higher in the winter.[13] The average July temperature is 55 °F (13 °C) in the morning and close to 90 °F (32 °C) in the afternoon. The average January temperature is 10 °F (−12 °C) in the morning and 40 °F (4 °C) in the afternoon, although variation between consecutive days can be 40 °F (22 °C).

West of the plains and foothills

View of the Western Slope from Grand Junction.

West of the plains and foothills, the weather of Colorado is much less uniform. Even places a few miles apart can experience entirely different weather depending on the topography of the area. Most valleys have a semi-arid climate, which becomes an alpine climate at higher elevations. Humid microclimates also exist in some areas. Generally, the wettest season in western Colorado is winter while June is the driest month. This is the opposite of precipitation patterns in the east. The mountains have cool summers with many days of high temperatures around 60 °F (16 °C) to 70 °F (21 °C), although frequent thunderstorms can cause sudden drops in temperature. Summer nights are cool or even cold at the highest elevations, which sometimes get snow even in the middle of the summer. The winters bring abundant, powdery snowfall to the mountains with plenty of sunshine in between major storms. The western slope has high summer temperatures similar to those found on the plains, while the winters tend to be slightly cooler due to the lack of warming winds common to the plains and Front Range. Other areas in the west have their own unique climate.

Extreme weather

Extreme weather is a common occurrence in Colorado. Thunderstorms are common east of the Continental divide in the spring and summer, and Colorado is one of the leading states in deaths due to lightning. Hail is a common sight in the mountains east of the divide and in the northwest part of the state. The Eastern Plains have some of the biggest hail storms in North America.[10] Also the Eastern Plains are part of Tornado Alley and produce some of the deadliest U.S. tornadoes. Some damaging tornadoes in the Eastern Plains include the 1990 Limon F3 tornado and the 2008 Windsor EF3 tornado, which devastated the town.[14] The plains are also susceptible to floods, which are caused both by thunderstorms and by the rapid melting of snow in the mountains during warm weather. Denver's record in 1921 for the number of consecutive days above 90 °F (32 °C) was broken during the summer of 2008. The new record of 24 consecutive days surpassed the previous record by almost a week.[15] Much of Colorado is a relatively dry state averaging only 17 inches (430 mm) of rain per year statewide and rarely experiences a time when some portion of the state is not in some degree of drought.[16] The lack of precipitation contributes to the severity of wildfires in the state such as the Hayman Fire, one of the largest wildfires in American history, and the Fourmile Canyon Fire of 2010, the most destructive wildfire in Colorado's recorded history.

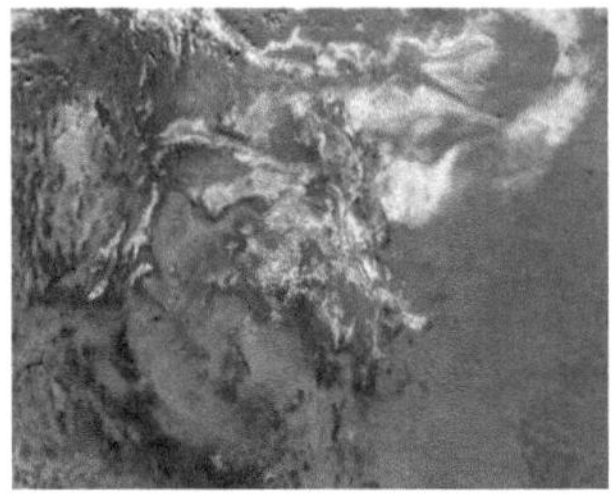

Snow highlights the rugged mountains as well as the urban and agricultural landscapes of the Colorado plains.

However, there are some of the mountainous regions of Colorado which receive a huge amount of moisture via winter snowfalls. The spring melts of these snows often cause great waterflows in such rivers as the Yampa River, the Grand River, the Colorado River, the Rio Grande, the Arkansas River, Cherry Creek, the North Platte River, and the South Platte River. Water flowing out of the Colorado Rocky Mountains is a very significant source of water for the farms, towns, and cities of fellow southwest states of New Mexico, Arizona, Utah, Nevada, as well as midwest like Nebraska and Kansas, and also southern states like Oklahoma and Texas. A significant amount of water is also diverted for use in California; occasionally (formerly naturally and consistently) the flow of water reaches northern Mexico.

Records

The highest temperature recorded in Colorado was 118 °F (48 °C) on July 11, 1888, at Bennett, whereas the lowest was −61 °F (−52 °C) on February 1, 1985, at Maybell.[17] [18]

Monthly normal high and low temperatures for various Colorado cities (°F)[19]

City	Jan	Feb	Mar	Apr	May	Jun	Jul	Aug	Sep	Oct	Nov	Dec
Alamosa	33/−4	40/5	50/16	59/23	68/32	78/40	82/46	79/45	72/36	62/24	46/11	35/−1
Colorado Springs	42/13	45/18	52/24	59/31	68/41	79/50	86/55	82/54	74/45	63/34	50/23	42/16
Denver	43/15	47/19	54/25	61/34	70/44	82/53	88/59	86/57	77/47	66/36	52/24	44/16
Grand Junction	37/16	45/23	56/31	64/38	74/46	87/55	96/61	92/60	83/50	67/39	50/26	39/18
Pueblo	45/14	50/19	57/26	65/34	75/45	86/54	91/59	89/58	81/49	69/35	54/22	45/15

Earthquakes

On August 22, 2011, a 5.3 magnitude earthquake occurred nine miles WSW of the city of Trinidad.[20] No casualties and only small damage was reported. It was the second largest earthquake in Colorado since a magnitude 5.7 earthquake was recorded in 1973.[21]

History

The region that is today the state of Colorado has been inhabited by Native Americans for more than 13 millennia. The Lindenmeier Site in Larimer County contains artifacts dating from approximately 11200 BCE to 3000 BCE. The eastern edge of the Rocky Mountains was a major migration route that was important to the spread of early peoples throughout the Americas. The Ancient Pueblo Peoples lived in the valleys and mesas of the Colorado Plateau.[22] The Ute Nation inhabited the mountain valleys of the Southern Rocky Mountains and the Western Rocky Mountains. The Arapaho Nation and the Cheyenne Nation moved west to hunt across the High Plains.

The ruins of the Cliff Palace at Mesa Verde as photographed by Gustaf Nordenskiöld in 1891.

The United States acquired a territorial claim to the eastern flank of the Rocky Mountains with the Louisiana Purchase from France in 1803. This U.S. claim conflicted with the claim of Spain to a huge region surrounding its colony of Santa Fé de Nuevo Méjico as its sovereign trading zone with native peoples. Zebulon Pike led a U.S. Army reconnaissance expedition into the disputed region in 1806. Colonel Pike and his men were arrested by Spanish cavalrymen in the San Luis Valley the following February, taken to Chihuahua, and then expelled from Mexico the following July.

The United States relinquished its claim to all land south and west of the Arkansas River and south of 42nd parallel north and west of the 100th meridian west as part of its purchase of Florida from Spain with the Adams-Onís Treaty of 1819. The treaty took effect February 22, 1821. Having settled its border with Spain, the United States admitted the southeastern portion of the Territory of Missouri to the Union as the state of Missouri on August 10, 1821. The remainder of the Missouri Territory, including what would become northeastern Colorado, became unorganized territory, and would remain so for 33 years over the question of slavery. After 11 years of war, Spain finally recognized the independence of Mexico with the Treaty of Córdoba signed on August 24, 1821. Mexico eventually ratified the Adams-Onís Treaty in 1831. The Texian Revolt of 1835–1836 fomented a dispute between the United States and Mexico which eventually erupted into the Mexican-American War in 1846. Mexico surrendered its northern territory to the United States with the Treaty of Guadalupe Hidalgo at the conclusion of the war in 1848.

Bent's Old Fort along the Arkansas River operated from 1833 to 1849.

Most American settlers traveling overland west to the Oregon Country, the new goldfields of California, or the new Mormon settlements of Deseret in the Salt Lake Valley, avoided the rugged Southern Rocky Mountains, and instead followed the North Platte River and Sweetwater River to South Pass, the lowest crossing of the Continental Divide between the Southern Rocky Mountains and the Central Rocky Mountains. In 1849, the Mormons of the Salt Lake Valley organized the extralegal State of Deseret, claiming the entire Great Basin and all lands drained by the Green, Grand, and Colorado rivers. The federal government of the United States flatly refused to recognize the new Mormon government, because it was theocratic and sanctioned plural marriage. Instead, the Compromise of 1850 divided the Mexican Cession and the northwestern claims of Texas into a new state and two new territories, the state of California, the Territory of New Mexico, and the Territory of Utah. On April 9, 1851, Mexican American settlers from the area of Taos settled the village of San Luis, then in the New Mexico Territory, later to become Colorado's first permanent Euro-American settlement.

In 1854, Senator Stephen A. Douglas persuaded the U.S. Congress to divide the unorganized territory east of the Continental Divide into two new organized territories, the Territory of Kansas and the Territory of Nebraska, and an unorganized southern region known as the Indian territory. Each new territory was to decide the fate of slavery within its boundaries, but this compromise merely served to fuel animosity between free soil and pro-slavery factions.

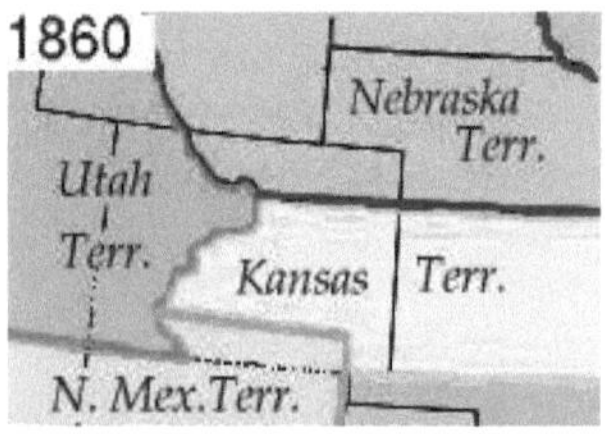

The territories of New Mexico, Utah, Kansas, and Nebraska before the creation of the Territory of Colorado.

Gold was discovered along the South Platte River then in western Kansas Territory in July 1858, precipitating the Pike's Peak Gold Rush.[23] The placer gold deposits along the rivers and streams of the region rapidly played out, but prospectors soon discovered far more valuable seams of hard rock gold in the nearby mountains.

The gold seekers organized the Provisional Government of the Territory of Jefferson on August 24, 1859, but this new territory failed to secure approval from the Congress of the United States embroiled in the debate over slavery. The election of Abraham Lincoln for the President of the United States on November 6, 1860, led to the secession of nine southern slave states and the threat of civil war among the states. Seeking to augment the political power of the Union states, the Republican Party dominated Congress quickly admitted the eastern portion of the Territory of Kansas into the Union as the free State of Kansas on January 29, 1861, leaving the western portion of the Kansas Territory, and its gold-mining areas, as unorganized territory.

A lithograph of the Denver City mining camp in 1859.

Thirty days later on February 28, 1861, outgoing U.S. President James Buchanan signed an Act of Congress organizing the free Territory of Colorado.[24] The original boundaries of Colorado remain unchanged today. The name Colorado was chosen because it was commonly believed that the Colorado River originated in the territory.[25] In 1776, Spanish priest Silvestre Vélez de Escalante recorded that Native Americans in the area knew the river as *el Rio Colorado* for the red-brown silt that the river carried from the mountains.[26] In 1859, a U.S. Army topographic expedition led by Captain John Macomb located the confluence of the Green River with the Grand River in what is now Canyonlands National Park in Utah.[27] The Macomb party designated the confluence as the source of the Colorado River.

The Battle of Glorieta Pass, 1862

On April 12, 1861, South Carolina artillery opened fire on Fort Sumter to start the American Civil War. While many gold seekers held sympathies for the Confederacy, the vast majority remained fiercely loyal to the Union cause. In 1862, a force of Texas cavalry invaded the Territory of New Mexico and captured Santa Fe on March 10. The object of this Western Campaign was to seize or disrupt the gold fields of Colorado and California and to seize ports on the Pacific Ocean for the Confederacy. A hastily organized force of Colorado volunteers force-marched from Denver City, Colorado Territory, to Glorieta Pass, New Mexico Territory, in an attempt to block the Texans. On March 28, the Coloradans and local New Mexico volunteers stopped the Texans at the Battle of Glorieta Pass, destroyed their cannon and supply wagons, and ran off 500 head of their horses and mules. The Texans were forced to retreat to Santa Fe. Having lost the supplies for their campaign and finding little support in New Mexico, the Texans abandoned Santa Fe and returned to San Antonio in defeat. The Confederacy made no further attempts to seize the Southwestern United States.

In 1864, Territorial Governor John Evans appointed the Reverend John Chivington as Colonel of the Colorado Volunteers with orders to protect white settlers from Cheyenne and Arapaho warriors who were accused of stealing cattle. Colonel Chivington ordered his men to attack a band of Cheyenne and Arapaho encamped along Sand Creek. Chivington reported that his troops killed more than 500 warriors. The militia returned to Denver City in triumph, but several officers reported that the so called battle was a blatant massacre of Indians at peace, that most of the dead were women and children, and that bodies of the dead had been mutilated and desecrated in hideous manner. Three U.S. Army inquiries condemned the action, and incoming President Andrew Johnson asked Governor Evans for his resignation, but none of the perpetrators was ever punished.

Mount of the Holy Cross was photographed by William Henry Jackson in 1874

In the midst and aftermath of Civil War, many discouraged prospectors returned to their homes, but a determined few stayed on to develop mines, mills, farms, ranches, roads, and towns in the Territory. On September 14, 1864, James Huff discovered silver near Argentine Pass, the first of many silver strikes. In 1867, the Union Pacific Railroad laid its tracks west to Weir, now Julesburg, in the northeast corner of the Territory. The Union Pacific linked up with the Central Pacific Railroad at Promontory Summit, Utah, on May 10, 1869, to form the First Transcontinental Railroad. The Denver Pacific Railway reached Denver in June of the following year, and the Kansas Pacific arrived two months later to forge the second line across the continent. In 1872, rich veins of silver were discovered in the San Juan Mountains on the Ute Indian reservation in southwestern Colorado. The Ute people were removed from the San Juans the following year.

The United States Congress passed an enabling act on March 3, 1875, specifying the requirements for the Territory of Colorado to become a state.[6] On August 1, 1876 (28 days after the Centennial of the United States), U.S. President Ulysses S. Grant signed a proclamation admitting Colorado to the Union as the 38th state and earning it the moniker "Centennial State".[28]

The discovery of a major silver lode near Leadville in 1878, triggered the Colorado Silver Boom. The Sherman Silver Purchase Act of 1890 invigorated silver mining, and Colorado's last, but greatest, gold strike at Cripple Creek a few months later lured a new generation of gold seekers. Colorado women were granted the right to vote beginning on November 7, 1893, making Colorado the second state to grant universal suffrage and the first one by a popular vote (of Colorado men). The repeal of the Sherman Silver Purchase Act in 1893 led to a staggering collapse of the mining and agricultural economy of Colorado, but the state slowly and steadily recovered.

The Georgetown Loop of the Colorado Central Railroad as photographed by William Henry Jackson in 1899

Colorado became the first western state to host a major political convention when the Democratic Party met in Denver in 1908. By the U.S. Census in 1930, the population of Colorado first exceeded one million residents. Colorado suffered greatly through the Great Depression and the Dust Bowl of the 1930s, but a major wave of immigration following World War II boosted Colorado's fortune. Tourism became a mainstay of the state economy, and high technology became an important economic engine. The United States Census Bureau estimated that the population of Colorado exceeded five million in 2009.

Three warships of the U.S. Navy have been named the USS *Colorado*. The first USS *Colorado* was named for the Colorado River. The later two ships were named in honor of the state, including the battleship USS *Colorado* which served in World War II in the Pacific beginning in 1941. At the time of the Attack on Pearl Harbor, this USS *Colorado* was located at the naval base in San Diego, Calif. and hence went unscathed.

Demographics

Historical populations		
Census	**Pop.**	**%±**
1860	34277	—
1870	39864	16.3%
1880	194327	387.5%
1890	413249	112.7%
1900	539700	30.6%
1910	799024	48.0%
1920	939629	17.6%
1930	1035791	10.2%
1940	1123296	8.4%
1950	1325089	18.0%
1960	1753947	32.4%
1970	2207259	25.8%
1980	2889964	30.9%
1990	3294394	14.0%
2000	4301262	30.6%
2010	5029196	16.9%
Sources: 1910-2010[29]		

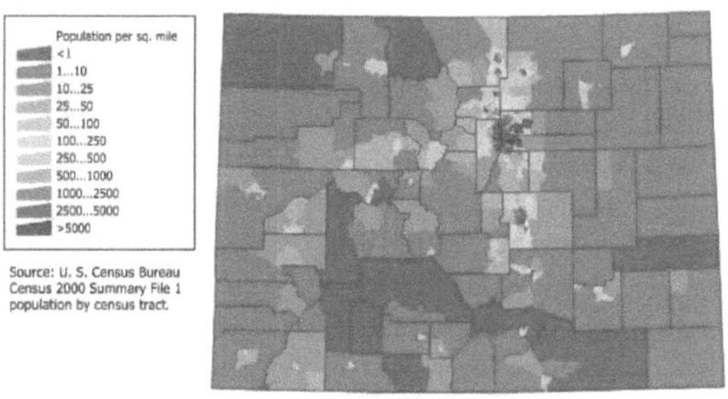

Colorado Population Density Map

Colorado's most populous city, and capital, is Denver. The Denver-Aurora-Boulder Combined Statistical Area with an estimated 2009 population of 3,110,436, is home to 61.90% of the state's residents.

As of 2005, Colorado has an estimated population of 4,665,177, which is an increase of 63,356, or 1.4%, from the prior year and an increase of 363,162, or 8.4%, since the year 2000. This includes a natural increase since the last census of 205,321 people (that is 353,091 births minus 147,770 deaths) and an increase due to net migration of 159,957 people into the state. Immigration from outside the United States resulted in a net increase of 112,217 people, and migration within the country produced a net increase of 47,740 people.

The largest increases are expected in the Front Range Urban Corridor, especially in the Denver metropolitan area. The state's fastest-growing counties are Douglas and Weld.[30] The center of population of Colorado is located just north of the village of Critchell in Jefferson County.[31]

Colorado has a high proportion of Hispanic, mostly Mexican-American, citizens in Metropolitan Denver, Colorado Springs, as well as the smaller cities of Greeley and Pueblo, and in many other smaller cities and towns all throughout the state. Colorado is well known for its strong Latino culture and presence. Southern, Southwestern, and Southeastern Colorado has a large number of Hispanos, the descendants of the early Mexican settlers of colonial Spanish origin. The 2000 United States Census found that 10.5% of people aged five and over in Colorado speak only Spanish at home, with the 2009 estimate being roughly 14%. Colorado also has a large immigration presence all throughout the state, which has led to Colorado cities being referred to as "Sanctuary Cities" for illegal immigrants as well. Colorado has the 5th highest percentage of undocumented people in the U.S., only behind Nevada, Arizona, California, and tied with Texas. An estimated 5.5-6.0% of the state's population is composed of illegal immigrants. Also, over 20% of the state's prisoners are undocumented inmates.[32] [33] Colorado, like New Mexico, is very rich in archaic Spanish idioms.[34]

Colorado also has some African-American communities located in Denver, in the neighborhoods of Montbello, Green Valley Ranch, Park Hill, Five Points, Whittier, and many other East Denver areas. A decent amount of African Americans are also found in Colorado Springs on the east and southeast side of the city. The state has sizable numbers of Asian-Americans of Mongolian, Chinese, Filipino, Korean, Southeast Asian and Japanese descent. The highest population of Asian Americans can be found on the south and southeast side of Denver, as well as some on Denver's southwest side. The Denver metropolitan area is considered more liberal and diverse than much of the state when it comes to political issues and environmental concerns.

According to the 2000 Census, the largest ancestry groups in Colorado are German (22%) including of Swiss and Austrian nationalities, Mexican (18%), Irish (12%), and English (12%). Persons reporting German ancestry are especially numerous in the Front Range, the Rockies (west-central counties) and Eastern parts/High Plains.[35] Denver, as well as all of Colorado, have numerous amount of predominately Latino neighborhoods and communities. Also, Denver and nearby areas on the Front Range has sizable German, Scandinavian, Italian, Slavic and Jewish American communities, partly a legacy of gold rushes in the late 19th century (1861–1889).

According to the 2010 U.S. Census, Colorado had a population of 5,029,196. In terms of race and ethnicity, the state was 81.3% White (70.0% Non-Hispanic White Alone), 4.0% Black or African American, 1.1% American Indian and Alaska Native, 2.8% Asian, 0.1% Native Hawaiian and Other Pacific Islander, 7.2% from Some Other Race, and 3.4% from Two or More Races. Hispanics and Latinos of any race made up 20.7% of the population.[36]

There were a total of 70,331 births in Colorado in 2006. (Birth Rate of 14.6). In 2007, non-Hispanic whites were involved in 59.1% of all the births.[37] Some 14.06% of those births involved a non-Hispanic white person and someone of a different race, most often with a couple including one Hispanic. A birth where at least one Hispanic person was involved counted for 43% of the births in Colorado.[38] As of the 2010 Census, Colorado has the seventh highest percentage of Hispanics (20.7%) in the U.S. behind New Mexico (46.3%), California (37.6%), Texas (37.6%), Arizona (29.6%), Nevada (26.5%), and Florida (22.5%). Per the 2000 census, the Hispanic population is estimated to be 918,899 or approximately 20% of the state total population. Colorado has the 4th largest population of Mexican-Americans behind California, Texas, and Arizona. In percentages, Colorado has the 6th highest percentage of Mexican-Americans behind New Mexico, California, Texas, Arizona, and Nevada.[39]

Religion

Major religious affiliations of the people of Colorado are:[40]

- Christian — 64%
  - Protestant — 44%
    - Evangelical — 23%
    - Mainline — 19%
    - Other Protestant — 2%
  - Roman Catholic — 19%
  - Latter Day Saint / Mormon — 2%
  - Orthodox — 1%
- Jewish — 2%
- Muslim — 1%
- Other Religions — 5%
- Unaffiliated — 25%

The Chapel on the Rock at Camp Saint Malo near Allenspark.

The Cadet Chapel at the United States Air Force Academy near Colorado Springs.

The largest denominations by number of adherents in 2000 were the Roman Catholic Church with 752,505; The Church of Jesus Christ of Latter-day Saints with 92,326 (133,727 year-end 2007) ; and Baptist with 85,083.[41]

At 25%, Colorado also has an above-average proportion of citizens who claim no religion. The U.S. average is 17%.

Health

Colorado also has a reputation for being a state of active and athletic people. According to several studies, Coloradans have the lowest rates of obesity of any state in the US.[42] As of 2007 the 18% of the population was considered medically obese, and while the lowest in the nation, the percentage had increased from 17% from 2004. Colorado Governor Bill Ritter commented: "As an avid fisherman and bike rider, I know first-hand that Colorado provides a great environment for active, healthy lifestyles," although he highlighted the need for continued education and support to slow the growth of obesity in the state.[43]

Culture

- List of museums in Colorado

Colorado History Museum in Denver

Fine arts

- Music of Colorado
- Theater in Colorado

Cuisine

Colorado is known for its Southwest and Rocky Mountain cuisine. Mexican restaurants are throughout the state.

Boulder, Colorado was named America's Foodiest Town 2010 by Bon Appétit.[44] Boulder, and Colorado in general, is home to a number of national food and beverage companies, top-tier restaurants and farmers' markets. Boulder, Colorado also has more Master Sommeliers per capita than any other city, including San Francisco and New York.[45]

The Food & Wine Classic held annually each June in Aspen, Colorado. Aspen also has a reputation as the culinary capital of the Rocky Mountain region.[46]

Denver, Colorado is known for steak, but now has a diverse culinary scene with many top-tier restaurants.[47]

Street art in Denver

Wine

Colorado wines include award-winning varietals that have attracted favorable notice from outside the state.[48] With wines made from traditional *Vitis vinifera* grapes along with wines made from cherries, peaches, plums and honey, Colorado wines have won top national and international awards for their quality.[49] Colorado's grape growing regions contain the highest elevation vineyards in the United States,[50] with most viticulture in the state practiced between 4000 feet (1219 m) and 7000 feet (2134 m) feet above sea level. The mountain climate ensures warm summer days and cool nights. Colorado is home to two designated American Viticultural Areas of the Grand Valley AVA and the West Elks AVA,[51] where most of the vineyards in the state are located. However, an increasing number of wineries are located along the Front Range.[52]

Economy

CNBC's list of "Top States for Business for 2010" has recognized Colorado as the third best state in the nation, falling short to only Texas and Virginia.[53]

Denver World Trade Center. The Denver financial district along 17th Street is known as the *Wall Street of the West.*

The United States quarter dollar coin released June 14, 2006, in honor of the state of Colorado.

The Bureau of Economic Analysis estimates that the total state product in 2010 was $257.6 billion.[54] Per capita personal income in 2010 was $51 940, ranking Colorado 11th in the nation.[55] The state's economy broadened from its mid-19th century roots in mining when irrigated agriculture developed, and by the late 19th century, raising livestock had become important. Early industry was based on the extraction and processing of minerals and agricultural products. Current agricultural products are cattle, wheat, dairy products, corn, and hay.

Corn growing in Larimer County

The federal government is also a major economic force in the state with many important federal facilities including NORAD, United States Air Force Academy and Peterson Air Force Base in Colorado Springs; NOAA and the National Institute of Standards and Technology in Boulder; U.S. Geological Survey and other government agencies at the Denver Federal Center in Lakewood; the Denver Mint, Buckley Air Force Base, and 10th Circuit Court of Appeals in Denver; and a federal Supermax Prison and other federal prisons near Cañon City. In addition to these and other federal agencies, Colorado has abundant National Forest land and four National Parks that contribute to federal ownership of 24615788 acres (99617 km^2) of land in Colorado, or 37% of the total area of the state.[56] In the second half of the 20th century, the industrial and service sectors have expanded greatly. The state's economy is diversified and is notable for its concentration of scientific research and high-technology industries. Other industries include food processing, transportation equipment, machinery, chemical products, the extraction of metals such as gold (see Gold mining in Colorado), silver, and molybdenum. Colorado now also has the largest annual production of beer of any state.[57] Denver is an important financial center.

Cattle ranching in Jackson County

An oil well in western Colorado

A number of nationally known brand names have originated in Colorado factories and laboratories. From Denver came the forerunner of telecommunications giant Qwest in 1879, Samsonite luggage in 1910, Gates belts and hoses in 1911, and Russell Stover Candies in 1923. Kuner canned vegetables began in Brighton in 1864. From Golden came Coors beer in 1873, CoorsTek industrial ceramics in 1920, and Jolly Rancher candy in 1949. CF&I railroad rails, wire, nails and pipe debuted in Pueblo in 1892. Holly Sugar was first milled from beets in Holly in 1905, and later moved its headquarters to Colorado Springs. The present-day Swift packed meat of Greeley evolved from Monfort of Colorado, Inc., established in 1930. Estes model rockets were launched in Penrose in 1958. Fort Collins has been the home of Woodward Governor Company's motor controllers (governors) since 1870, and Waterpik dental water jets and showerheads since 1962. Celestial Seasonings herbal teas have been made in Boulder since 1969. Rocky Mountain Chocolate Factory made its first candy in Durango in 1981.

Colorado has a flat 4.63% income tax, regardless of income level. Unlike most states, which calculate taxes based on federal *adjusted gross income*, Colorado taxes are based on *taxable income* – income after federal exemptions and federal itemized (or standard) deductions.[58] [59] Colorado's state sales tax is 2.9% on retail sales. When state revenues exceed state constitutional limits, full-year Colorado residents can claim a sales tax refund on their individual state income tax return. Many counties and cities charge their own rates in addition to the base state rate. There are also certain county and special district taxes that may apply.

Real estate and personal business property are taxable in Colorado. The state's senior property tax exemption was temporarily suspended by the Colorado Legislature in 2003. The tax break is scheduled to return for assessment year 2006, payable in 2007.

As of September 2010, the state's unemployment rate is 8.2%.[60]

Philanthropy

Major philanthropic organizations based in Colorado, including the Daniels Fund, the Anschutz Family Foundation, the Gates Family Foundation, the El Pomar Foundation and the Boettcher Foundation, grant approximately $400 million[61] each year from approximately $7 billion[62] of assets.

Natural resources

Colorado has significant hydrocarbon resources. According to the Energy Information Administration, Colorado hosts seven of the Nation's 100 largest natural gas fields and two of its 100 largest oil fields. Conventional and unconventional natural gas output from several Colorado basins typically account for more than 5 percent of annual U.S. natural gas production. Colorado's oil shale deposits hold an estimated 1 trillion barrels (160 km^3) of oil – nearly as much oil as the entire world's proven oil reserves; the economic viability of the oil shale, however, has not been demonstrated.[63] Substantial deposits of bituminous, subbituminous, and lignite coal are found in the state. Kimberlite volcanic pipes have been found in Colorado; the Kelsey Lake Diamond Mine operated for several years, recovering gem quality diamonds.

Colorado's high Rocky Mountain ridges and eastern plains offer wind power potential, and geologic activity in the mountain areas provides potential for geothermal power development. Much of the state is sunny and could produce solar power. Major rivers flowing from the Rocky Mountains offer hydroelectric power resources. Corn grown in the flat eastern part of the state offers potential resources for ethanol production.

Transportation

Colorado transportation system connects its cities, residents, and visitors through a diverse and well regulated modes.

Highways and roads

Colorado state welcome sign

Interstate and Federal Highway Routes in Colorado

- Interstate 25
- Interstate 70
- Interstate 76
- Interstate 225
- Interstate 270
- U.S. Route 6
- U.S. Route 24
- U.S. Route 34
- U.S. Route 36
- U.S. Route 40
- U.S. Route 50
- U.S. Route 84
- U.S. Route 85
- U.S. Route 87
- U.S. Route 138
- U.S. Route 160
- U.S. Route 285
- U.S. Route 287
- U.S. Route 350
- U.S. Route 400
- U.S. Route 491
- U.S. Route 550

Aviation

Denver International Airport

Commercial Airports in Colorado

- ALS – San Luis Valley Regional Airport
- ASE – Aspen-Pitkin County Airport
- CEZ – Cortez Municipal Airport
- COS – City of Colorado Springs Municipal Airport
- **DEN – Denver International Airport**[64]
- DRO – Durango-La Plata County Airport
- EGE – Eagle County Regional Airport
- FNL – Fort Collins-Loveland Municipal Airport
- GJT – Grand Junction Regional Airport
- GUC – Gunnison-Crested Butte Regional Airport
- HDN – Yampa Valley Airport
- MTJ – Montrose Regional Airport
- PUB – Pueblo Memorial Airport
- TEX – Telluride Regional Airport
- WBU - Boulder Municipal Airport

Rail

Union Station in Denver.

Amtrak Passenger Railroad Routes through Colorado

- *California Zephyr:* Chicago, Galesburg, Omaha, **Fort Morgan, Denver, Winter Park, Granby, Glenwood Springs, Grand Junction**, Salt Lake City, Reno, Sacramento, and Emeryville
- *Southwest Chief:* Chicago, Galesburg, Kansas City, Topeka, **Lamar, La Junta, Trinidad**, Raton, Lamy, Albuquerque, Gallup, Flagstaff, and Los Angeles

Mass transit

Communities in Colorado with Regional Bus Service

Alamosa, Aurora, Boulder, Brush, Berthoud, **Colorado Springs**, Delta, **Denver**, Durango, Englewood, Frisco, **Fort Collins**, Fort Morgan, Glenwood Springs, **Grand Junction**, **Greeley**, Lamar, Limon, Longmont, Loveland, Montrose, **Pueblo**, Rocky Ford, Springfield, Sterling, Trinidad, Vail, and Walsenburg

- Denver and Colorado Springs are serviced by the FREX regional bus.
- Fort Collins, Loveland, Berthoud, and Longmont are serviced by the FLEX regional bus.[65]

Accidents

In 2011, Colorado ranked among the five deadliest states for debris/litter–caused vehicle accidents per total number of registered vehicles and population size. Figures derived from the National Highway Traffic Safety Administration (NHTSA) show at least 16 persons in Colorado were killed each year in motor vehicle collisions with non-fixed objects, including debris, dumped litter, animals and their carcasses.

In the United States including Colorado, most civil aviation incidents are investigated by the National Transportation Safety Board (NTSB), as well as the Colorado Bureau of Investigations (CBI). When investigating an aviation disaster, NTSB investigators piece together evidence from the crash and determine the likely cause or causes. Where as the CBI will also investigate if there is any involved criminal actions.

Government and politics

State government

Gubernatorial election results

Year	Republican	Democratic
2010	36.4% *651,232*	**51.0%** *912,005*
2006	40.16% *625,886*	**56.98%** *888,096*
2002	**62.62%** *884,584*	33.65% *475,373*
1998	**49.06%** *648,202*	48.43% *639,905*
1994	38.70% *432,042*	**55.47%** *619,205*
1990	35.43% *358,403*	**61.89%** *626,032*

The Colorado State Capitol in Denver

Just like all the states, Colorado's state constitution provides for three branches of government: the legislative, the executive, and the judicial branches. The Governor heads the state's executive branch. The Colorado Supreme Court is the highest judicial court in the state. The state legislative body is the Colorado General Assembly, which is made up of two houses, the House of Representatives and the Senate.

The House has 65 members and the Senate has 35. Currently, the House is controlled by the Republican Party by a one vote majority and the Senate is controlled by the Democratic Party. The 2005 Colorado General Assembly was the first to be controlled by the Democrats in forty years. The incumbent governor is Democrat John Hickenlooper.

Most Coloradans are originally native to other states (nearly 60% according to the 2000 census),[66] and this is illustrated by the fact that the state did not have a native-born governor from 1975 (when John David Vanderhoof left office) until 2007, when Bill Ritter took office; his election the previous year marked the first electoral victory for a native-born Coloradan in a gubernatorial race since 1958 (Vanderhoof had ascended from the Lieutenant Governorship when John Arthur Love was given a position in Richard Nixon's administration in 1973).

Federal politics

Presidential elections results

Year	Republican	Democratic
2008	44.71% *1,073,584*	**53.66%** *1,288,568*
2004	**51.69%** *1,101,255*	47.02% *1,001,732*
2000	**50.75%** *883,745*	42.39% *738,227*
1996	**45.80%** *691,848*	44.43% *671,152*
1992	35.87% *562,850*	**40.13%** 629,681
1988	**53.06%** *728,177*	45.28% *621,453*

Colorado is considered a swing state in both state and federal elections. Coloradans have elected 17 Democrats and 12 Republicans to the governorship in the last 100 years. In presidential politics, Colorado supported Democrats Bill Clinton in 1992 and Barack Obama in 2008, and supported Republicans Robert J. Dole in 1996 and George W. Bush in 2000 and 2004. The presidential outcome in 2008 was the second closest to the national popular vote, after Virginia.[67]

Colorado politics has the contrast of conservative cities such as Colorado Springs and liberal cities such as Boulder. Democrats are strongest in metropolitan Denver, the college towns of Fort Collins and Boulder, southern Colorado (including Pueblo), and a few western ski resort counties. The Republicans are strongest in the Eastern Plains, Colorado Springs, Greeley, and far Western Colorado near Grand Junction.

Former Colorado senator and attorney general Ken Salazar is the current United States Secretary of the Interior (as of January 20, 2009).

The state of Colorado is represented by its two United States Senators:

- United States Senate Class 2 – Mark Emery Udall (Democratic) 2009–
- United States Senate Class 3 – Michael Farrand Bennet (Democratic)[68] 2009–

Colorado is represented by seven Representatives to the United States House of Representatives:

- Colorado's 1st congressional district - Diana Louise DeGette (Democratic) 1997-
- Colorado's 2nd congressional district - Jared Schutz Polis (Democratic) 2009-
- Colorado's 3rd congressional district - Scott Tipton (Republican) 2011–
- Colorado's 4th congressional district - Cory Gardner (Republican) 2011–
- Colorado's 5th congressional district - Douglas L. "Doug" Lamborn (Republican) 2007-
- Colorado's 6th congressional district - Michael "Mike" Coffman (Republican) 2009-
- Colorado's 7th congressional district - Edwin George "Ed" Perlmutter (Democratic) 2007-

Counties

The state of Colorado is divided into 64 counties.[69] Counties are important units of government in Colorado since the state has no secondary civil subdivisions such as townships. Two of these counties, the City and County of Denver and the City and County of Broomfield, have consolidated city and county governments.

Nine Colorado counties have a population in excess of 250,000 each, while eight Colorado counties have a population of less than 2,500 each. The ten most populous Colorado counties are located in the Front Range Urban Corridor.

An enlargeable map of the 64 counties of the state of Colorado

The 25 Most Populous Colorado Counties 2010

Pop Rank	County	2010 Census	2000 Census	Pop Change
1	El Paso County	622263	516929	+20.38%
2	Denver City and County	600158	554636	+8.21%
3	Arapahoe County	572003	487967	+17.22%
4	Jefferson County[70]	534543	527056	+1.42%
5	Adams County[70]	441603	363857	+21.37%
6	Larimer County	299630	251494	+19.14%
7	Boulder County[70]	294567	291288	+1.13%
8	Douglas County	285465	175766	+62.41%
9	Weld County[70]	252825	180936	+39.73%
10	Pueblo County	159063	141472	+12.43%
11	Mesa County	146723	116255	+26.21%
12	Garfield County	56389	43791	+28.77%
13	Broomfield City and County[70]	55889	0	NA
14	Eagle County	52197	41659	+25.30%
15	La Plata County	51334	43941	+16.82%
16	Fremont County	46824	46145	+1.47%
17	Montrose County	41276	33432	+23.46%
18	Delta County	30952	27834	+11.20%

19	Morgan County	28159	27171	+3.64%
20	Summit County	27994	23548	+18.88%
21	Montezuma County	25535	23830	+7.15%
22	Routt County	23509	19690	+19.40%
23	Teller County	23350	20555	+13.60%
24	Elbert County	23086	19872	+16.17%
25	Logan County	22709	20504	+10.75%

Metropolitan areas

The United States Office of Management and Budget (OMB) has defined one Combined Statistical Area (CSA),[71] seven Metropolitan Statistical Areas (MSAs),[72] and seven Micropolitan Statistical Areas (μSAs)[73] in the state of Colorado.[74]

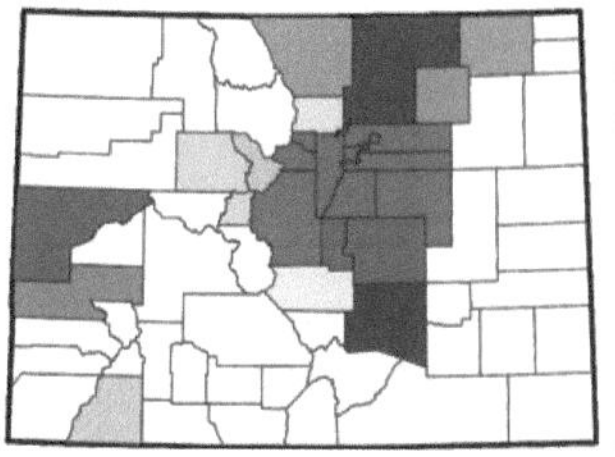

Map of the 14 Core Based Statistical Areas in the state of Colorado.

The most populous of the 14 Core Based Statistical Areas in Colorado is the Denver-Aurora-Broomfield, CO Metropolitan Statistical Area. This area had a population of 2,543,482 at the 2010 United States Census, an increase of 17.88% since the 2000 United States Census.[75]

The more extensive Denver-Aurora-Boulder, CO Combined Statistical Area had a population of 3,090,874 at the 2010 United States Census, an increase of +17.52% since the 2000 United States Census.[75]

The most populous extended metropolitan region in Rocky Mountain Region is the Front Range Urban Corridor along the northeast face of the Southern Rocky Mountains. This region with Denver at its center had a population of 4,333,742 at the 2010 United States Census, an increase of +17.50% since the 2000 United States Census.[75]

Municipalities

The state of Colorado currently has 271 active incorporated municipalities, including 196 towns, 73 cities, and two consolidated city and county governments.[76] [77]

Colorado municipalities operate under one of five types of municipal governing authority. Colorado has one town with a territorial charter, 160 statutory towns, 12 statutory cities, 96 home rule municipalities (61 cities and 35 towns), and 2 consolidated city and county governments.

The skyline of downtown Denver with Speer Boulevard in the foreground

The 25 Most Populous Colorado Municipalities 2010

Pop Rank	Municipality	2010 Census	2000 Census	Pop Change
1	City and County of Denver	600158	554636	+8.21%
2	City of Colorado Springs	416427	360890	+15.39%
3	City of Aurora	325078	276393	+17.61%
4	City of Fort Collins	143986	118652	+21.35%
5	City of Lakewood	142980	144126	−0.80%
6	City of Thornton	118772	82384	+44.17%
7	City of Pueblo	106595	102121	+4.38%
8	City of Arvada	106433	102153	+4.19%
9	City of Westminster	106114	100940	+5.13%
10	City of Centennial[78]	100377	0	NA
11	City of Boulder	97385	94673	+2.86%
12	City of Greeley	92889	76930	+20.74%
13	City of Longmont	86270	71093	+21.35%
14	City of Loveland	66859	50608	+32.11%
15	City of Grand Junction	58566	41986	+39.49%
16	City and County of Broomfield[70]	55889	38272	+46.03%
17	Town of Castle Rock	48231	20224	+138.48%
18	City of Commerce City	45913	20991	+118.73%
19	Town of Parker	45297	23558	+92.28%
20	City of Littleton	41737	40340	+3.46%
21	City of Northglenn	35789	31575	+13.35%
22	City of Brighton	33352	20905	+59.54%
23	City of Englewood	30255	31727	−4.64%
24	City of Wheat Ridge	30166	32913	−8.35%
25	City of Fountain	25846	15197	+70.07%

The City and County of Denver and the City of Aurora both levy an Occupational Privilege Tax (OPT or Head Tax) on employers and employees.

- If any employee performs work in the city limits and is paid over US$500.00 for that work in a single month, the employee and employer are both liable for the OPT regardless of where the main business office is located or headquartered.
- In Denver, the employer is liable for US$4.00 per employee per month and the employee is liable for US$5.75 per month.
- In Aurora, both employer and employees are liable for US$2.00 per month.
- It is the employer's responsibility to withhold, remit, and file the OPT returns. If an employer does not comply, they can be held liable for both portions of the OPT as well as penalties and interest.

Unincorporated communities

In addition to its 271 municipalities, Colorado has 187 unincorporated United States census designated places and many other small communities.

The 15 Most Populous Census Designated Places in Colorado 2010

Pop Rank	Census Designated Place	2010 Census	2000 Census	Pop Change
1	Highlands Ranch	96713	70931	+36.35%
2	Security-Widefield	32882	29845	+10.18%
3	Ken Caryl	32438	30887	+5.02%
4	Dakota Ridge	32005	NA	NA
5	Pueblo West	29637	16899	+75.38%
6	Columbine	24280	24095	+0.77%
7	Clifton	19889	17345	+14.67%
8	Sherrelwood	18287	17657	+3.57%
9	Cimarron Hills	16161	15194	+6.36%
10	Welby	14846	12973	+14.44%
11	Fort Carson	13813	10566	+30.73%
12	Black Forest	13116	13247	−0.99%
13	Berkley	11207	10743	+4.32%
14	Cherry Creek	11120	NA	NA
15	The Pinery	10517	7253	+45.00%

Districts

The state of Colorado has more than 3,000 districts with taxing authority. These districts may provide schools, law enforcement, fire protection, water, sewage, drainage, irrigation, transportation, recreation, infrastructure, cultural facilities, business support, redevelopment, or other services.

Some of these districts have authority to levy sales tax and well as property tax and use fees. This has led to a hodgepodge of sales tax and property tax rates in Colorado. There are some street intersections in Colorado with a different sales tax rate on each corner, sometimes substantially different.

Some of the more notable Colorado districts are:

- The Regional Transportation District (RTD), which affects the counties of Denver, Boulder, Jefferson, and portions of Adams, Arapahoe, Broomfield, and Douglas Counties
- The Scientific and Cultural Facilities District (SCFD), a special regional tax district with physical boundaries contiguous with county boundaries of Adams, Arapahoe, Boulder, Broomfield, Denver, Douglas, and Jefferson Counties
  - It is a 0.1% retail sales and use tax (one penny on every $10).
  - According to the Colorado statute, the SCFD distributes the money to local organizations on an annual basis. These organizations must provide for the enlightenment and entertainment of the public through the production, presentation, exhibition, advancement or preservation of art, music, theater, dance, zoology, botany, natural history or cultural history.
  - As directed by statute, SCFD recipient organizations are currently divided into three "tiers" among which receipts are allocated by percentage.

    - Tier I includes regional organizations: the Denver Art Museum, the Denver Botanic Gardens, the Denver Museum of Nature and Science, the Denver Zoo, and the Denver Center for the Performing Arts. It receives 65.5%.
    - Tier II currently includes 26 regional organizations. Tier II receives 21%.
    - Tier III has over 280 local organizations such as small theaters, orchestras, art centers, and natural history, cultural history, and community groups. Tier III organizations apply for funding to the county cultural councils via a grant process. This tier receives 13.5%.
  - An 11-member board of directors oversees the distributions in accordance with the Colorado Revised Statutes. Seven board members are appointed by county commissioners (in Denver, the Denver City Council) and four members are appointed by the Governor of Colorado.
- The Football Stadium District (FD or FTBL), approved by the voters to pay for and help build the Denver Broncos' stadium INVESCO Field at Mile High
- Local Improvement Districts (LID) within designated areas of southeast Jefferson and Boulder counties
- Regional Transportation Authority (RTA) taxes at varying rates in Basalt, Carbondale, Glenwood Springs, Gunnison County

Education

Colleges and universities in Colorado:

The United States Air Force Academy.

The University of Colorado at Boulder.

The University of Denver.

- Adams State College
- Aims Community College
- Arapahoe Community College
- Art Institute of Colorado
- Belleview College
- Colorado Christian University
- Colorado College
- Colorado Mesa University
- Colorado Mountain College
- Colorado Northwestern Community College
- Colorado School of Mines
- Metropolitan State College of Denver
- Morgan Community College
- Naropa University
- National Technological University
- Nazarene Bible College
- Northeastern Junior College
- Otero Junior College
- Pikes Peak Community College
- Pueblo Community College
- Red Rocks Community College
- Redstone College

- Colorado State University System
  - Colorado State University
  - Colorado State University-Pueblo
- Colorado Technical University
- Community College of Aurora
- Community College of Denver
- Denver Seminary
- DeVry University
- Emily Griffith Opportunity School
- Fort Lewis College
- Front Range Community College
- Iliff School of Theology
- Johnson & Wales University
- Jones International University
- Lamar Community College
- Regis University
- Rocky Mountain College of Art and Design
- Rocky Vista University College of Osteopathic Medicine
- Trinidad State Junior College
- United States Air Force Academy
- University of Colorado System
  - University of Colorado at Boulder
  - University of Colorado at Colorado Springs
  - University of Colorado Denver
    - Anschutz Medical Campus
    - Auraria Campus
- University of Denver
- University of Northern Colorado
- Western State College

Military installations

- Air Reserve Personnel Center
- Buckley Air Force Base
- Fort Carson (U.S. Army)
  - Piñon Canyon Maneuver Site
- Peterson Air Force Base
  - Cheyenne Mountain Directorate
- Pueblo Chemical Depot (U.S. Army)
- Schriever Air Force Base
- United States Air Force Academy

Peterson Air Force Base

Protected areas

Units of the National Park System in Colorado:

- Arapaho National Recreation Area[79]
- Bent's Old Fort National Historic Site
- Black Canyon of the Gunnison National Park
- Canyons of the Ancients National Monument[80]
- Colorado National Monument
- Continental Divide National Scenic Trail[81]
- Curecanti National Recreation Area
- Dinosaur National Monument
- Florissant Fossil Beds National Monument
- Great Sand Dunes National Park and Preserve
- Hovenweep National Monument
- Mesa Verde National Park and UNESCO World Heritage Site

Great Sand Dunes National Park and Preserve

- Old Spanish National Historic Trail
- Pony Express National Historic Trail
- Rocky Mountain National Park
- Sand Creek Massacre National Historic Site
- Santa Fe National Historic Trail
- Yucca House National Monument

Spruce Tree House in Mesa Verde National Park

Sports

Professional sports teams

Colorado is the least populous state with a franchise in each of the major professional sports leagues.

The Colorado Rockies National League baseball club at Coors Field in Denver.

Sports Authority Field at Mile High, home of the Denver Broncos National Football League club and the Denver Outlaws Major League Lacrosse club.

Pepsi Center in Denver, home of the Denver Nuggets National Basketball Association club, the Colorado Avalanche National Hockey League club, and the Colorado Mammoth National Lacrosse League club.

Professional Sports Clubs based in Colorado

Club	Home	First game	Sport	League
Denver Broncos	Denver	September 9, 1960	Football	National Football League
Denver Barbarians	Denver	Spring 1967	Rugby Union	Rugby Super League
Denver Nuggets	Denver	September 27, 1967	Basketball	National Basketball Association
Colorado Springs Sky Sox	Colorado Springs	June 18, 1988	Baseball	Minor League Baseball (AAA)
Colorado Rockies	Denver	April 5, 1993	Baseball	Major League Baseball
Colorado Avalanche	Denver	October 6, 1995	Ice hockey	National Hockey League
Colorado Rapids	Commerce City	April 13, 1996	Soccer	Major League Soccer
Colorado Mammoth	Denver	January 3, 2003	Lacrosse	National Lacrosse League
Colorado Eagles	Loveland	October 17, 2003	Ice hockey	ECHL
Denver Outlaws	Denver	May 20, 2006	Lacrosse	Major League Lacrosse

College athletics

The following universities and colleges participate in the National Collegiate Athletic Association Division I.

NCAA Division I athletic programs in Colorado

Team	School	City	Conference
Air Force Falcons	United States Air Force Academy	Colorado Springs	Mountain West
Colorado Buffaloes	University of Colorado at Boulder	Boulder	Pac 12
Colorado College Tigers	Colorado College	Colorado Springs	WCHA[82]
Colorado State Rams	Colorado State University	Fort Collins	Mountain West
Denver Pioneers	University of Denver	Denver	WCHA[82] and Sun Belt
Northern Colorado Bears	University of Northern Colorado	Greeley	Big Sky

See also

- List of federal lands in Colorado
- List of National Register of Historic Places in Colorado
- List of people from Colorado
- List of places in Colorado
- Mountain peaks of Colorado
- U.S. state

References

[1] "Elevations and Distances in the United States" (http://erg.usgs.gov/isb/pubs/booklets/elvadist/elvadist.html#Highest). U.S. Geological Survey. April 29, 2005. . Retrieved October 19, 2007.

[2] "National Geodetic Survey data sheet KL0637 for Mount Elbert" (http://www.ngs.noaa.gov/cgi-bin/ds_mark.prl?PidBox=KL0637). National Geodetic Survey. . Retrieved October 19, 2007.

[3] "Colorado - Definition" (http://www.merriam-webster.com/dictionary/colorado?show=0&t=1307303426). Merriam-webster.com. 2010-08-13. . Retrieved 2011-06-05.

[4] Writers Style Guide (http://ccs.colostate.edu/style_guide/style_guide_az.aspx#state_names), Colorado State University. Retrieved January 19, 2009.

[5] Quillen, Ed (2007-03-18). "Coloradoan or Coloradan" (http://www.denverpost.com/opinion/ci_5447358). Denverpost.com. . Retrieved 2010-07-30.

[6] Forty-third United States Congress (March 3, 1875). "An Act to Enable the People of Colorado to Form a Constitution and State Government, and for the Admission of the Said State into the Union on an Equal Footing with the Original States" (http://web.archive.org/web/20080507151300/http://www.i2i.org/Publications/ColoradoConstitution/cnenable.htm) (PDF). Archived from the original (http://www.i2i.org/Publications/ColoradoConstitution/cnenable.htm) on 2008-05-07. . Retrieved April 14, 2008.

[7] http://maps.google.com/?ie=UTF8&ll=38.258671,-108.873138&spn=0.731113,1.235962&z=10

[8] "Rectangular States and Kinky Borders" (http://www.maa.org/mathtourist/mathtourist_08_30_07.html). Maa.org. 2007-08-30. . Retrieved 2010-07-30.

[9] U.S. Geological Survey. "Elevations and Distances" (http://erg.usgs.gov/isb/pubs/booklets/elvadist/elvadist.html#Highest). . Retrieved September 8, 2006.

[10] Doesken, Nolan J.; Roger A. Pielke, Sr., Odilia A.P. Bliss (January 2003). "Climate of Colorado" (http://ccc.atmos.colostate.edu/climateofcolorado.php). Colorado Climate Center – Department of Atmospheric Science – Colorado State University. . Retrieved January 25, 2009.

[11] U.S. Forest Service. "Rocky Mountain Region 14ers" (http://www.fs.fed.us/r2/recreation/14ers/). . Retrieved November 6, 2009.

[12] "Pikes Peak, Colorado" (http://www.peakbagger.com/peak.aspx?pid=5689). Peakbagger.com. . Retrieved October 10, 2009.

[13] DRI.edu (http://www.wrcc.dri.edu/narratives/COLORADO.htm) Western Regional Climate Center. Retrieved October 24, 2006.

[14] Slater, Jane (May 28, 2008). "Thursday's Tornado State's 4th Costliest Disaster" (http://www.thedenverchannel.com/news/16408116/detail.html). KMGH. .

[15] "Denver's Consecutive 90 Degree Streaks" (http://www.crh.noaa.gov/bou/?n=consec90). National Weather Service. . Retrieved October 10, 2009.

[16] "A History of Drought" (http://ccc.atmos.colostate.edu/pdfs/ahistoryofdrought.pdf) (PDF). . Retrieved 2010-07-30.

[17] "Record Highest Temperatures by State" (http://lwf.ncdc.noaa.gov/oa/pub/data/special/maxtemps.pdf) (PDF). National Climatic Data Center. January 1, 2004. . Retrieved January 11, 2007.

[18] "Record Lowest Temperatures by State" (http://lwf.ncdc.noaa.gov/oa/pub/data/special/mintemps.pdf) (PDF). National Climatic Data Center. January 1, 2004. . Retrieved January 11, 2007.

[19] UStravelweather.com (http://www.ustravelweather.com/weather-colorado/)

[20] http://earthquake.usgs.gov/earthquakes/recenteqsus/Maps/US2/36.38.-106.-104.php

[21] "Largest Colorado quake since 1973 shakes homes" (http://www.usatoday.com/news/nation/story/2011-08-23/Largest-Colorado-quake-since-1973-shakes-homes/50102314/1?csp=34news&utm_source=feedburner&utm_medium=feed&utm_campaign=Feed:+usatoday-NewsTopStories+(News+-+Top+Stories)). *USA Today*. August 23, 2011. . Retrieved August 23, 2011.

[22] " Genocide Wiped Out Native American Population (http://news.discovery.com/archaeology/genocide-native-americans-ethnic-cleansing.html)", Discovery News, September 20, 2010.

[23] Gehling, Richard (2006). "The Pike's Peak Gold Rush" (http://web.archive.org/web/20080628172253/http://www.geocities.com/Heartland/Falls/2000/index.html). Richard Gehling. Archived from the original (http://www.geocities.com/Heartland/Falls/2000/index.html) on 2008-06-28. . Retrieved June 12, 2007.

[24] "An Act to provide a temporary Government for the Territory of Colorado" (http://www.colorado.gov/dpa/doit/archives/territory.pdf) (PDF). Thirty-sixth United States Congress. February 28, 1861. . Retrieved June 12, 2007.

[25] Early explorers identified the Gunnison River in Colorado as the headwaters of the Colorado River. The Grand River in Colorado was later tentively identified as the primary headwaters of the river. Finally in 1916, E.C. LaRue, the Chief Hydrologist of the United States Geological Survey, identified the Green River in southwestern Wyoming as the proper headwaters of the actual, overall Colorado River.
[26] State of Colorado – Division of Information Technologies. "State Names and Nicknames" (http://www.colorado.gov/dpa/doit/archives/history/symbemb.htm#Name). . Retrieved November 15, 2006.
[27] Report of the exploring expedition from Santa Fé, New Mexico, to the junction of the Grand and Green Rivers of the great Colorado of the West, in 1859: under the command of Capt. J. N. Macomb, Corps of topographical engineers, Volume 1 @ archive.org (http://www.archive.org/stream/reportofexplorin00unitrich#page/n5/mode/2up)
[28] President of the United States of America (August 1, 1876). "Proclamation of the Admission of Colorado to the Union" (http://www.presidency.ucsb.edu/ws/index.php?pid=70540) (php). The American Presidency Project. . Retrieved April 14, 2008.
[29] Resident Population Data. "Resident Population Data - 2010 Census" (http://2010.census.gov/2010census/data/apportionment-pop-text.php). 2010.census.gov. . Retrieved 2011-06-05.
[30] "Population growth - Colorado counties" (http://www.epodunk.com/top10/countyPop/coPop6.html). Epodunk.com. . Retrieved 2011-06-05.
[31] "Population and Population Centers by State – 2000" (http://www.census.gov/geo/www/cenpop/statecenters.txt). United States Census Bureau. . Retrieved December 4, 2008.
[32] Fred Elbel, CAIR (2005-05-19). "Crime and illegal aliens in Colorado - CAIR - Colorado Alliance for Immigration Reform" (http://www.cairco.org/issues/issues_crime_colorado.html). Cairco.org. . Retrieved 2011-06-05.
[33] "Language Map Data Center" (http://www.mla.org/map_data_results&state_id=8&mode=state_tops&order=r). Mla.org. 2007-07-17. . Retrieved 2010-07-30.
[34] "talking about Colorado in "nada"" (http://www.elcastellano.org/palabra.php?id=946). Elcastellano.org. 2007-06-30. . Retrieved 2010-07-30.
[35] US. "Map of Latitude: 39.500656 Longitude: -105.203628, by MapQuest" (http://www.mapquest.com/maps/map.adp?searchtype=address&formtype=address&latlongtype=decimal&latitude=39.500656&longitude=-105.203628). Mapquest.com. . Retrieved 2010-07-30.
[36] http://factfinder2.census.gov/faces/tableservices/jsf/pages/productview.xhtml?pid=DEC_10_DP_DPDP1&prodType=table
[37] "National Vital Statistics Reports, Volume 57, Number 12, (3/18/2009)" (http://www.cdc.gov/nchs/data/nvsr/nvsr57/nvsr57_12.pdf) (PDF). . Retrieved 2010-07-30.
[38] CDPHE.state.co.us (http://www.cdphe.state.co.us/scripts/htmsql.exe/cohid/natalityPub.hsql), COHID Birth Data Request
[39] (http://www.statemaster.com/state/CO-colorado/peo-people), Statemaster Colorado
[40] "U.S. Religion Map and Religious Populations – U.S. Religious Landscape Study – Pew Forum on Religion & Public Life" (http://religions.pewforum.org/maps). Religions.pewforum.org. . Retrieved 2010-07-30.
[41] "The Association of Religion Data Archives | Maps & Reports" (http://www.thearda.com/mapsReports/reports/state/08_2000.asp). Thearda.com. . Retrieved 2011-06-05.
[42] http://calorielab.com/news/wp-images/post-images/fattest-states-2007-big.gif
[43] "Fattest States 2010: CalorieLab's Annual Obesity Map - State Obesity Rankings | CalorieLab - Health News & Information Blog" (http://calorielab.com/news/2007/08/06/fattest-states-2007/). CalorieLab. 2010-06-28. . Retrieved 2011-06-05.
[44] Knowlton, Andrew. "America's Foodiest Town 2010: Boulder, Colorado: In the Magazine" (http://www.bonappetit.com/magazine/2010/10/americas_foodiest_town_2010_boulder). bonappetit.com. . Retrieved 2011-06-05.
[45] (http://www.denvermagazine.com/March-2011/Colorado-039s-Master-Sommeliers/)
[46] Arnold, Katie (2008-06-08). "As Skiers Depart Aspen, Chowhounds Take Their Place" (http://travel.nytimes.com/2008/06/08/travel/08Choice.html). Aspen (Colo): Travel.nytimes.com. . Retrieved 2011-06-05.
[47] "Colorado Travel Guide" (http://www.travelandleisure.com/guides/colorado). Travelandleisure.com. . Retrieved 2011-06-05.
[48] Jackenthal, Stefani (2008-10-05). "Biking Colorado's Wine Country" (http://travel.nytimes.com/2008/10/05/travel/05biking.html). Colorado: Travel.nytimes.com. . Retrieved 2011-06-05.
[49] "The Jefferson Cup Invitational Wine Competition" (http://www.thejeffersoncup.com). Thejeffersoncup.com. 2010-11-24. . Retrieved 2011-06-05.
[50] "Wine Industry Feature Articles - Is Colorado the New Washington?" (http://www.winesandvines.com/template.cfm?section=features&content=48734). Winesandvines.com. . Retrieved 2011-06-05.
[51] http://www.ttb.gov/appellation/us_by_ava.pdf
[52] "Colorado Wine Industry Development Board" (http://www.coloradowine.com/overview.html). Coloradowine.com. . Retrieved 2011-06-05.
[53] America's Top States for Business 2010." CNBC Special Report (2010): 1. Web. 9 May 2011. <http://www.cnbc.com/id/37516043/>.
[54] "GDP by State" (http://greyhill.com/gdp-by-state). Greyhill Advisors. . Retrieved 7 September 2011.
[55] "References" (http://www.bea.gov/newsreleases/regional/spi/2009/pdf/spi0309.pdf) (PDF). . Retrieved 2010-07-30.
[56] Tony Frank (January 1997). "Colorado Land Ownership by County (acres)" (http://web.archive.org/web/20060116094720/http://www.ag.state.co.us/resource/colorado_land_ownership.html) (Excel). Colorado Department of Agriculture. Archived from the original (http://www.ag.state.co.us/resource/documents/TotalCOLandOwnership.xls) on January 16, 2006. . Retrieved July 15, 2007. Colorado Department of Agriculture: Land Ownership (http://www.ag.state.co.us/resource/colorado_land_ownership.html)

[57] "Colorado rides on Fat Tire to beer heights" (http://www.rockymountainnews.com/news/2007/nov/24/reuteman-colorado-rides-on-fat-tire-to-beer/). Rockymountainnews.com. 2007-11-24. . Retrieved 2011-06-05.

[58] Colorado individual income tax return (2005) Revenue.state.co.us (http://www.revenue.state.co.us/PDF/05104f.pdf). Retrieved September 26, 2006.

[59] U.S. Individual Income Tax Return (2005) online copy (http://www.irs.gov/pub/irs-pdf/f1040.pdf#search="form 1040 site:irs.gov"). Retrieved September 26, 2006.

[60] Bls.gov (http://www.bls.gov/lau/); Local Area Unemployment Statistics

[61] Coloradofunders.org (http://www.coloradofunders.org/docs/2007 Giving Study PublishedFinal_1.pdf)

[62] http://www.cof.org/files/Documents/Government/StateGiving/CO.pdf

[63] "EIA State Energy Profiles: Colorado" (http://tonto.eia.doe.gov/state/state_energy_profiles.cfm?sid=CO). June 12, 2008. . Retrieved June 24, 2008.

[64] Denver International Airport was the ninth busiest airport on Earth in 2006.

[65] "FLEX bus: Longmont, Berthoud, Loveland, Ft. Collins get connected" (http://biggreenboulder.com/environment/flex-bus-longmont-berthoud-loveland-fort-collins/). BigGreenBoulder. 2010-06-07. . Retrieved 2011-06-05.

[66] "State of Residence in 2000 by State of Birth" (http://www.census.gov/population/www/cen2000/briefs/phc-t38/index.html). US Census Bureau. . Retrieved October 10, 2009.

[67] Goodman, Josh. "Move Over Missouri, Iowa Is the New Bellwether State" (http://ballotbox.governing.com/2008/11/move-over-missouri-iowa-is-the-bellwether-state.html). Governing.com. . Retrieved October 10, 2009.

[68] Governor Bill Ritter appointed Michael Bennet to serve the remaining two years of United States Senator Ken Salazar term of office which was left vacant on January 20, 2009, when new United States President Barack Obama appointed the Colorado Senator to serve in his Cabinet as United States Secretary of the Interior.

[69] "Colorado Counties" (http://www.dola.state.co.us/dlg/local_governments/counties.html). State of Colorado, Department of Local Affairs. January 8, 2007. . Retrieved January 30, 2007.

[70] The City and County of Broomfield, Colorado was created on November 15, 2001, from portions of Boulder County, Adams County, Jefferson County, and Weld County, Colorado.

[71] The United States Office of Management and Budget defines a Combined Statistical Area (CSA) as an aggregate of adjacent Core Based Statistical Areas that are linked by commuting ties.

[72] The United States Office of Management and Budget defines a Metropolitan Statistical Area (MSA) as a Core Based Statistical Area having at least one urbanized area of 50,000 or more population, plus adjacent territory that has a high degree of social and economic integration with the core as measured by commuting ties.

[73] The United States Office of Management and Budget defines a Micropolitan Statistical Area (μSA) as a Core Based Statistical Area having at least one urban cluster of at least 10,000 but less than 50,000 population, plus adjacent territory that has a high degree of social and economic integration with the core as measured by commuting ties.

[74] "OMB Bulletin No. 10-02: Update of Statistical Area Definitions and Guidance on Their Uses" (http://www.whitehouse.gov/omb/assets/bulletins/b10-02.pdf). United States Office of Management and Budget. December 1, 2009. . Retrieved January 7, 2010.

[75] "American Factfinder" (http://factfinder2.census.gov/main.html). United States Census Bureau. . Retrieved May 10, 2011.

[76] "Active Colorado Municipalities" (http://www.dola.state.co.us/dlg/local_governments/municipalities.html). State of Colorado, Department of Local Affairs. . Retrieved May 11, 2011.

[77] "Colorado Local Government by Type" (http://www.dola.state.co.us/dlg/local_governments/lgtypes.html). State of Colorado, Department of Local Affairs. February 27, 2007. . Retrieved May 11, 2011.

[78] The City of Centennial, Colorado was incorporated on February 7, 2001

[79] Managed by the United States Department of Agriculture, National Forest Service.

[80] Managed by the United States Department of the Interior, Bureau of Land Management.

[81] Jointly managed by the United States Department of Agriculture, National Forest Service, and the United States Department of the Interior, National Park Service and Bureau of Land Management.

[82] The Western Collegiate Hockey Association is a National Collegiate Athletic Association Division I ice hockey-only conference.

Further reading

- Explore Colorado, A Naturalist's Handbook, The Denver Museum of Natural History and Westcliff Publishers, 1995, ISBN 1-56579-124-X for an excellent guide to the ecological regions of Colorado.
- The Archeology of Colorado, Revised Edition, E. Steve Cassells, Johnson Books, Boulder, Colorado, 1997, trade paperback, ISBN 1-55566-193-9.
- Chokecherry Places, Essays from the High Plains, Merrill Gilfillan, Johnson Press, Boulder, Colorado, trade paperback, ISBN 1-55566-227-7.
- The Tie That Binds, Kent Haruf, 1984, hardcover, ISBN 0-03-071979-8, a fictional account of farming in Colorado.
- Railroads of Colorado: Your Guide to Colorado's Historic Trains and Railway Sites, Claude Wiatrowski, Voyageur Press, 2002, hardcover, 160 pages, ISBN 0-89658-591-3

External links

State government

- Colorado state government website (http://www.Colorado.gov/)
  - Colorado information (http://www.colorado.gov/cs/Satellite/CO-Portal/CXP/1174084099742)
  - Colorado state government (http://www.colorado.gov/cs/Satellite/CO-Portal/CXP/1165693060247)
  - Colorado state agencies (http://www.colorado.gov/cs/Satellite/CO-Portal/CXP/1184337612761)
    - Colorado Department of Transportation (http://www.dot.state.co.us/)
      - Colorado highway maps (http://www.dot.state.co.us/App_DTD_DataAccess/Maps/index.cfm?fuseaction=MapsMain&MenuType=Maps)
        - Colorado Travel Map (http://dtdexternal.dot.state.co.us/travelmap/)
  - Colorado counties (http://dola.colorado.gov/dlg/local_governments/counties.html)
  - Colorado municipalities (http://dola.colorado.gov/dlg/local_governments/municipalities.html)
  - Colorado special districts (http://dola.colorado.gov/dlg/local_governments/lgov_a.html)
  - Colorado tourism (http://www.colorado.com/)
- Colorado State and County Government Websites (http://www.n2genealogy.com/colorado.html#government)
- List of searchable databases produced by Colorado state agencies (http://wikis.ala.org/godort/index.php/Colorado) hosted by the American Library Association Government Documents Roundtable (http://wikis.ala.org/godort/index.php/Main_Page).

Federal government

- Energy & Environmental Data for Colorado (http://tonto.eia.doe.gov/state/state_energy_profiles.cfm?sid=CO)
- USGS Colorado state facts, real-time, geographic, and other scientific resources of Colorado (http://www.usgs.gov/state/state.asp?State=CO)
- United States Census Bureau (http://www.census.gov/)
  - Colorado QuickFacts (http://quickfacts.census.gov/qfd/states/08000.html)
  - Colorado economic data (http://factfinder.census.gov/servlet/QTTable?_bm=n&_lang=en&qr_name=DEC_2000_SF3_U_DP3&ds_name=DEC_2000_SF3_U&geo_id=04000US08)
  - Colorado housing data (http://factfinder.census.gov/servlet/QTTable?_bm=n&_lang=en&qr_name=DEC_2000_SF3_U_DP4&ds_name=DEC_2000_SF3_U&geo_id=04000US08)
  - Colorado social data (http://factfinder.census.gov/servlet/QTTable?_bm=n&_lang=en&qr_name=DEC_2000_SF3_U_DP2&ds_name=DEC_2000_SF3_U&geo_id=04000US08)
- USDA ERS Colorado state facts (http://www.ers.usda.gov/statefacts/co.htm)

Other

- Colorado travel guide from Wikitravel
- Colorado County Evolution (http://www.stanwyck.com/cogenweb/cocounties.html)
- Ask Colorado (http://www.askcolorado.org/)
- 2000 Census of Population and Housing for Colorado (http://www2.census.gov/census_2000/datasets/demographic_profile/Colorado/2kh08.pdf)
- Mountain and Desert Plants of Colorado and the Southwest (http://swcoloradowildflowers.com/),
- Climate of Colorado (http://www.wrcc.dri.edu/narratives/COLORADO.htm)
- Colorado (http://www.dmoz.org/Regional/North_America/United_States/Colorado/) at the Open Directory Project

gag:Kolorado mrj:Колорадо

Republican City, Nebraska

Republican City, Nebraska	
— Village —	
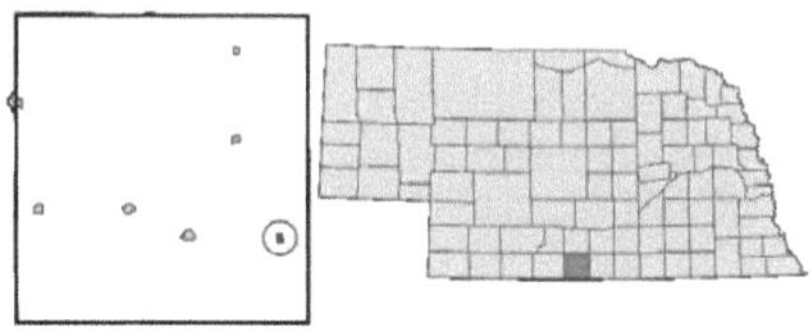 Location of Republican City, Nebraska	
Coordinates: 40°5′52″N 99°13′17″W	
Country	United States
State	Nebraska
County	Harlan
Area	
- Total	0.3 sq mi (0.9 km^2)
- Land	0.3 sq mi (0.9 km^2)
- Water	0.0 sq mi (0.0 km^2)
Elevation	2028 ft (618 m)
Population (2000)	
- Total	209
- Density	633.8/sq mi (244.7/km^2)
Time zone	Central (CST) (UTC-6)
- Summer (DST)	CDT (UTC-5)
ZIP code	68971
Area code(s)	308
FIPS code	31-41130[1]
GNIS feature ID	0835013[2]

Republican City is a village in Harlan County, Nebraska, United States. The population was 209 at the 2000 census. The town is near the Harlan County Reservoir.

Geography

Republican City is located at 40°5′52″N 99°13′17″W (40.097708, -99.221301)[3] .

According to the United States Census Bureau, the village has a total area of 0.3 square miles (0.78 km^2), all of it land.

Demographics

As of the census[1] of 2000, there were 209 people, 107 households, and 69 families residing in the village. The population density was 633.8 people per square mile (244.5/km²). There were 218 housing units at an average density of 661.1 per square mile (255.1/km²). The racial makeup of the village was 99.52% White, and 0.48% from two or more races. Hispanic or Latino of any race were 0.96% of the population.

There were 107 households out of which 15.0% had children under the age of 18 living with them, 57.9% were married couples living together, 4.7% had a female householder with no husband present, and 35.5% were non-families. 33.6% of all households were made up of individuals and 20.6% had someone living alone who was 65 years of age or older. The average household size was 1.95 and the average family size was 2.43.

In the village the population was spread out with 12.9% under the age of 18, 5.7% from 18 to 24, 11.0% from 25 to 44, 34.0% from 45 to 64, and 36.4% who were 65 years of age or older. The median age was 60 years. For every 100 females there were 101.0 males. For every 100 females age 18 and over, there were 100.0 males.

The median income for a household in the village was $26,250, and the median income for a family was $32,500. Males had a median income of $27,500 versus $17,917 for females. The per capita income for the village was $14,863. About 6.9% of families and 8.5% of the population were below the poverty line, including 10.5% of those under the age of eighteen and 12.2% of those sixty five or over.

References

[1] "American FactFinder" (http://factfinder.census.gov). United States Census Bureau. . Retrieved 2008-01-31.

[2] "US Board on Geographic Names" (http://geonames.usgs.gov). United States Geological Survey. 2007-10-25. . Retrieved 2008-01-31.

[3] "US Gazetteer files: 2010, 2000, and 1990" (http://www.census.gov/geo/www/gazetteer/gazette.html). United States Census Bureau. 2011-02-12. . Retrieved 2011-04-23.

Alma, Nebraska

Alma, Nebraska	
— City —	
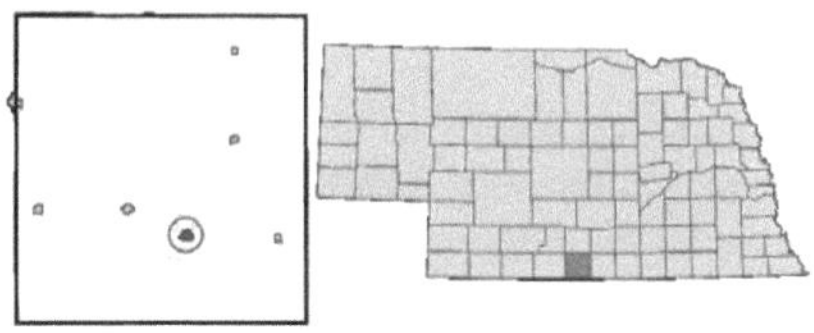 Location of Alma, Nebraska	
Coordinates: 40°6′3″N 99°21′45″W	
Country	United States
State	Nebraska
County	Harlan
Government	
- Mayor	Hal Haeker
Area	
- Total	0.7 sq mi (1.8 km^2)
- Land	0.7 sq mi (1.8 km^2)
- Water	0.0 sq mi (0.0 km^2)
Elevation	1975 ft (602 m)
Population (2000)	
- Total	1214
- Density	1744.7/sq mi (673.6/km^2)
Time zone	Central (CST) (UTC-6)
- Summer (DST)	CDT (UTC-5)
ZIP code	68920
Area code(s)	308
FIPS code	31-00975[1]
GNIS feature ID	0826967[2]

Alma is a city in Harlan County, Nebraska, United States. The population was 1,214 at the 2000 census. It is the county seat of Harlan County[3]. The town is named after a daughter of one of the early settlers. The town is near the Harlan County Reservoir.

Geography

Alma is located at 40°6′3″N 99°21′45″W (40.100733, -99.362618)[4] .

According to the United States Census Bureau, the city has a total area of 0.7 square miles (1.8 km^2), all of it land.

Alma is located at the junction of US Route 136 and US Route 183. The community is at the northwest end of Harlan County Lake.

Demographics

As of the census[1] of 2000, there were 1,214 people, 520 households, and 318 families residing in the city. The population density was 1,744.7 people per square mile (669.6/km²). There were 594 housing units at an average density of 853.7 per square mile (327.6/km²). The racial makeup of the city was 99.18% White, 0.08% Native American, 0.08% Asian, 0.08% from other races, and 0.58% from two or more races. Hispanic or Latino of any race were 0.58% of the population.

There were 520 households out of which 25.6% had children under the age of 18 living with them, 54.0% were married couples living together, 5.0% had a female householder with no husband present, and 38.8% were non-families. 36.3% of all households were made up of individuals and 22.7% had someone living alone who was 65 years of age or older. The average household size was 2.23 and the average family size was 2.93.

In the city the population was spread out with 24.3% under the age of 18, 3.4% from 18 to 24, 19.8% from 25 to 44, 23.9% from 45 to 64, and 28.7% who were 65 years of age or older. The median age was 46 years. For every 100 females there were 83.9 males. For every 100 females age 18 and over, there were 82.3 males.

The median income for a household in the city was $27,315, and the median income for a family was $36,250. Males had a median income of $26,786 versus $17,065 for females. The per capita income for the city was $14,795. About 9.5% of families and 13.1% of the population were below the poverty line, including 17.5% of those under age 18 and 10.0% of those age 65 or over.

History

Alma was founded in the spring of 1871 by a group of Union Pacific railroad laborers originally from Cheyenne, Wyoming. Harlan County was created by the Nebraska Legislature on June 3, 1871, and Alma was made the county seat on July 3, 1871. A dispute among the original settlers led to a dissolution of the town, but it was re-organized in 1874.

Alma's status as county seat was challenged by residents of Orleans, but an 1884 Nebraska Supreme Court decision affirmed Alma's status as seat of Harlan County. Alma officially incorporated in July 1881.

On January 13, 1880 the Burlington Railroad reached Alma, providing a significant boost for local commerce and industry. In 1887, the Kansas City and Omaha Railroad also reached Alma. Telephone service reached Alma in June 1899. Electricity arrived with a steam power plant in 1906 and water service followed in 1907.

In 1935, major flooding of the nearby Republican River provided an impetus to build a dam on the river. The Army Corps of Engineers started construction of the Harlan County Dam on August 1, 1946, and completed work in November 1952. The new artificial lake, Harlan County Lake, became one of Alma's top recreation destinations.

During World War II, Alma hosted a small number of captured German soldiers.

References

[1] "American FactFinder" (http://factfinder.census.gov). United States Census Bureau. . Retrieved 2008-01-31.

[2] "US Board on Geographic Names" (http://geonames.usgs.gov). United States Geological Survey. 2007-10-25. . Retrieved 2008-01-31.

[3] "Find a County" (http://www.naco.org/Counties/Pages/FindACounty.aspx). National Association of Counties. . Retrieved 2011-06-07.

[4] "US Gazetteer files: 2010, 2000, and 1990" (http://www.census.gov/geo/www/gazetteer/gazette.html). United States Census Bureau. 2011-02-12. . Retrieved 2011-04-23.

- Flasnick, Bob, editor (1992). *Alma: Towards the Year 2000.* Harlan County Journal.

External links

- Official city website (http://www.almacity.com/)
- NPPD website for Alma (http://www.ci.alma.ne.us/)
- Harlan County Lake (http://www.harlanlake.com/)
- Hoesch Memorial Public Library (http://www.almalibrary.com/)
- University of Nebraska-Lincoln Extension in Harlan County (http://www.harlan.unl.edu/)

Brownville, Nebraska

Brownville, Nebraska	
— Village —	

Carson House on Main Street, in Brownville Historic District

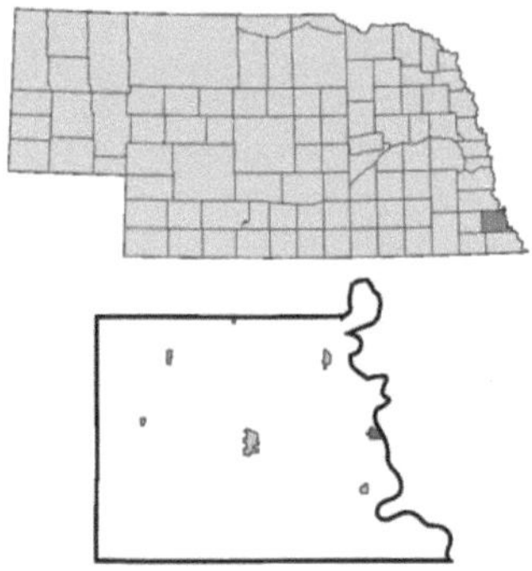

Location of Brownville, Nebraska

Coordinates: 40°23′49″N 95°39′34″W	
Country	United States
State	Nebraska
County	Nemaha
Area	
- Total	0.6 sq mi (1.6 km^2)
- Land	0.6 sq mi (1.6 km^2)
- Water	0.0 sq mi (0.0 km^2)
Elevation	958 ft (292 m)
Population (2000)	
- Total	146

- Density	233.9/sq mi (90.3/km^2)
Time zone	Central (CST) (UTC-6)
- Summer (DST)	CDT (UTC-5)
ZIP code	68321
Area code(s)	402
FIPS code	31-06750[1]
GNIS feature ID	0827698[2]

Brownville is a village in Nemaha County, Nebraska, United States. The population was 146 at the 2000 census.

History

Established in 1854 and incorporated in 1856, Brownville was the largest town in the Nebraska Territory, with a population of 1,309 by 1880. Bordering slave-holding Missouri, the town became an important port on the Missouri River. Daniel Freeman, the first homesteader to file a claim under the Homestead Act of 1862, staked his claim at a New Year's Eve party in Brownville.

The rise of the railroad was ultimately Brownville's undoing. The railroads siphoned traffic away from the Missouri River's steamboats. Brownville's attempt to secure a railroad of its own was severely botched and led to immense tax increases to pay the bonds for the failed venture. This drove most of the population away and led to the county seat being transferred to Auburn in 1885.[3] In 1856 Thomas Weston Tipton moved to the town to serve as a minister. He later became a longtime U.S. Senator, and ran an unsuccessful campaign for Governor of Nebraska. Omaha land magnates Herman and Augustus Kountze held large holdings in the early town.

After the Civil War David Mercer established a law practice in the town. Mercer later served as a long-time U.S. Representative from Nebraska. The Nebraska State Fair was held in the town in 1870 and 71. That year Eugene Gilmore, the future acting Governor-General of the Philippines, was born in the town.

The Brownville Bridge was built over the Missouri River in 1939. Today it is on the National Register of Historic Places. Owing to its place in Nebraska history, Brownville is primarily a tourist attraction with several old houses opened for tours or converted into museums. Art galleries and wineries in and near Brownville have also helped to make the town an increasingly attractive getaway for residents of the region's larger cities. Omaha's *River City Star* was built in the town in 1967, and was originally named the *Belle of Brownville*. The Governor Furnas Arboretum was planted in the city in 1992. The arboretum is named in honor of Nebraska's second governor, Robert W. Furnas, the signer of the first declaration of Arbor Day. The Cooper Nuclear Station is south of the city. In 1995 a mild tornado blew through the town.

Geography

According to the United States Census Bureau, Brownville has a total area of 0.6 square miles (1.6 km^2), all of it land.

Notable places

Brownville sits in the Loess Hills above the Missouri River Valley. It is home to one of Nebraska's two nuclear power plants. The Cooper Nuclear Station is owned and operated by the Nebraska Public Power District. U.S. Route 136, which meets Nebraska Highway 67 near Brownville, runs through the town, exiting the state via the Brownville Bridge. At Brownville bicycle riders along the Lewis & Clark Trail Bicycle Route can choose between the Steamboat Trace or the Hamburg Option.

Demographics

As of the census[1] of 2000, there were 146 people, 74 households, and 40 families residing in the village. The population density was 233.9 people per square mile (90.9/km²). There were 106 housing units at an average density of 169.8 per square mile (66.0/km²). The racial makeup of the village was 99.32% White, and 0.68% from two or more races.

There were 74 households out of which 20.3% had children under the age of 18 living with them, 44.6% were married couples living together, 5.4% had a female householder with no husband present, and 44.6% were non-families. 40.5% of all households were made up of individuals and 17.6% had someone living alone who was 65 years of age or older. The average household size was 1.97 and the average family size was 2.63.

In the village the population was spread out with 18.5% under the age of 18, 5.5% from 18 to 24, 19.2% from 25 to 44, 34.2% from 45 to 64, and 22.6% who were 65 years of age or older. The median age was 51 years. For every 100 females there were 117.9 males. For every 100 females age 18 and over, there were 108.8 males.

The median income for a household in the village was $34,375, and the median income for a family was $51,000. Males had a median income of $35,750 versus $28,438 for females. The per capita income for the village was $20,928. There were 11.5% of families and 15.9% of the population living below the poverty line, including 15.2% of under eighteens and 17.9% of those over 64.

Notable residents

- Thomas Tipton, one of the first two United States Senators from Nebraska.
- Eugene Allen Gilmore, Acting Governor-General of the Philippines in 1927.

Cultural references

- Willa Cather's 1897 short story *A Resurrection* is set in Brownville, and it includes a description of the town.
- *St. Joseph, Days of Civil War*, a documentary about the Civil War, was partly filmed in Brownville.

References

[1] "American FactFinder" (http://factfinder.census.gov). United States Census Bureau. . Retrieved 2008-01-31.
[2] "US Board on Geographic Names" (http://geonames.usgs.gov). United States Geological Survey. 2007-10-25. . Retrieved 2008-01-31.
[3] Brownville's 150th Anniversary 1854 - 2004: The Village of Firsts Remains on the Frontier Edge (http://www.yourcountryneighbor.com/YCN/BHS150.pdf). Captain Meriwether Lewis Museum. Retrieved 12/13/08.

External links

- Official town website (http://www.brownville-ne.com/)
- Comprehensive website (http://www.brownvillemills-ne.com/village/)

Article Sources and Contributors

Harlan County Reservoir *Source*: http://en.wikipedia.org/w/index.php?title=Harlan_County_Reservoir *Contributors*: Aaroncj, BD2412, Backspace, D6, DandyDan2007, Docu, Freechild, Hmains, J04n, JaGa, Jllm06, Lightmouse, Matt Deres, Pakaraki, Tassedethe, Woohookitty

Nebraska *Source*: http://en.wikipedia.org/w/index.php?title=Nebraska *Contributors*: 16@r, 1brettsnyder, 2004-12-29T22:45Z, 223fms, 2829NE, 5 albert square, 90873e, A m b buko, ANSWER 12, Aaron north, AbsolutDan, Ackerman elementry, Addiction2Code, Addshore, Ahassan05, Ahoerstemeier, Aioth, Aitias, Aivazovsky, Akradecki, Alansohn, Ale jrb, Aleccolgan, Alex.muller, AlexiusHoratius, Alexwcovington, Ali, AlphaEta, Alphachimp, Alvinrune, Amadeus2347, Amazonien, Amikeco, Ammodramus, Amog, Andrew Weevlos, Andy Marchbanks, Andy120290, Angr, Antandrus, AntonioJackson, Appraiser, Arrowcatcher, AssistantX, Astrocommander, Astynax, AuburnPilot, Avicennasis, Awd1970, Azazello, BSveen, Babylonian Armor, Babysage, Backstop21, Baguasquirrel, Baiji, Balshan, Barmej, Barneca, Barticus88, Bazonka, Beannaithe, Beginning, Beland, Belligero, Belovedfreak, BenBaker, Bender235, Berean Hunter, Bhuenink, Bigmike1985, Billwhittaker, BirgitteSB, Biruitorul, Bjoh249, Bkell, Bkonrad, Blanchardb, Blathnaid, Blehfu, Blueboy96, Bluedogtn, BobKawanaka, Bobo192, Bomac, Bongwarrior, BoomMan77, BoomerAB, BornintheUSA, Brain40, Brainiack16, Brainyiscool, BrendelSignature, Brett37, Brickawaye, Brion VIBBER, Brockert, Buaidh, Buckeyetigre, Bullshark44, Bullyoffrogs, Bumm13, Butwhatdoiknow, CSWarren, CUTKD, CWY2190, Caiaffa, Caj8, Caltas, Can't sleep, clown will eat me, CanadianLinuxUser, Caponer, CaptainJackWill, Carlaude, Carter, Casper2k3, Catdude, Cburnett, Cchow2, ChrisRuvolo, Chrislk02, ChristTrekker, Christy747, Chun-hian, Civil Engineer III, Civilian knowledge, Ckatz, Ckincaid77, ClairSamoht, Clarityfiend, Closedmouth, Cmichael, Cnmerrill01, Cntras, Colinsvo, Colleenthegreat, Colonies Chris, CommonsDelinker, Conversion script, Coulraphobic123, Courcelles, CowboyJoe771, Cpl Syx, Craigmartin49, CrazyC83, CrazyChemGuy, Cremepuff222, Cribbswh, Crosbiesmith, Cst17, Czj, D6, DC, DCEdwards1966, DJ-Siren, Daa89563, Dabomb87, Dale Arnett, DandyDan2007, Danieldaglish, Danny, Danwebb12, Darkwind, Daven200520, Davish Krail, Gold Five, Db099221, Dcornwall, DeadEyeArrow, Dean52, Deflective, Deidara8293, Deli nk, Dellpie, DerHexer, Derim Hunt, Diego Grez, Dimadick, Dinopup, Dionyseus, Discospinster, Djsasso, Dkdantastic, Dmartin969, Doc glasgow, Doczilla, Dogs2000, Donstr, Doubleplusjeff, Doulos Christos, Dr. Blofeld, Dr.Farshad, Dragonsrage92, Drkeithphd, Drmies, Dsnow75, Dude1818, Dufekin, Dwayne, Dyanega, Dysprosia, ERcheck, Eagle4000, Eco84, Ed g2s, Edgar h8ter, Edivorce, Eeekster, Egrimaldo, Ellywa, Emeraldcityserendipity, EncycloPetey, Enviroboy, Epbr123, Epistemophiliac, Epolk, ErgoSum88, Ethanlong9, Eu.stefan, Everyking, Evice, Excirial, Expertiseinc, Eóin, Farmersgranddaughter, Feezo, Fishing, Flatterworld, Flayshar, Flyguy649, Foundationperson, Frank101, Fredbauder, Freechild, Frenchwhale, Frog47, From Selma to Fresno, Fry1989, Ftc08, Fyyer, GB fan, GSlicer, Gail, Gaius Cornelius, Gene Nygaard, Gilliam, Gimmetrow, Gogo Dodo, Gohiking, Golgofrinchian, Good Olfactory, Grammarmonger, Grammarxxx, GreenRunner0, GregU, Gsmgm, Gurch, Gwalla, HalfShadow, Harmil, HarveyHobgoblin, Haukurth, Hdt83, Hintswen, Hmains, Hobartimus, HolmesJohnC, Hombrepollo, Hotgirl1096, Hstrotman, Hu12, Hurtitle, Hushpuckena, I SKIN GAYS, IRP, Icairns, Imarookie, Inbierno, Indianbiscut, Intelligentsium, Intellinet, Irde altz, Iridescent, IronGargoyle, Ironman 222, Ixfd64, J.delanoy, JForget, JR98664, JaGa, Jamescole94, Jamieamberbuie, Jasynnash2, Jauhienij, JeanneLPD, Jedikaiti, Jeff G., Jengirl1988, Jengod, Jeremiestrother, Jfitts, Jguad1, Jhendin, JimIrwin, Jimglazer, JinJian, Jjcoolguy, Joe Schmedley, John K, John254, Jojhutton, JonClee86, Jonathan Kovaciny, Jose77, Joseph1245, Jphillips66, Jsayre64, Judeeclare, Juliancolton, Jusdafax, Jusjih, JustinHagstrom, KJS77, Kansan, Kblack24, Kerryboy1, KevinCuddeback, Kintetsubuffalo, Klperson, Knoodles21, Kostisl, Koyaanis Qatsi, Kozuch, Kurowoofwoof111, Kwamikagami, L Kensington, LairepoNite, Landon1980, Larkinp, LcawteHuggle, Levineps, Lhfeditor, Liblamb, LightCMM, Lightmouse, Lilac Soul, Lily20, Linuxerist, Lisa Dawne Kelly, LittleDan, LokiClock, Looxix, Lord Emsworth, Lord Voldemort, Luk, M C Y 1008, MJCdetroit, MJHankel, Macintosh User, Madhero88, Magosta, Malthazar, Maltmomma, Manuel Trujillo Berges, Maritabarita, Martpol, Mateinsixtynine, Mattgibney, Mattsalt77, Mav, Maximillion Pegasus, Maximini1010, Maximus Rex, Mayfly may fly, Mcelite, Mckito, Mdben, Mendaliv, Mentifisto, Mermaid from the Baltic Sea, Merman, Migozared, MikeGruz, Mild Bill Hiccup, Miles530, Miq, Miranda, Miskwito, MonoAV, Mooremoore, Moptopsrock, Moverton, Mr Stephen, Mr. Lefty, MrChupon, Mshecket, Mtownene, Mu Cow, Myanw, Myrealana, Myrhtrips, N-k, NEICenergy, Naalyax3sg, Nakon, Nauticashades, NebraskaDawg, Nebraskagirl123, Netoholic, Neutrality, Neverquick, Nezzadar, Nickj 47, Nikki88, Nintendonurd, Nishkid64, Nn123645, Noahkoch, NorCalHistory, NorsemanII, Nphs2010, Oda Mari, Ohnoitsjamie, Okiefromokla, Omahapubliclibrary, Oo64eva, Ordtoy, Owain, Oxymoron83, Pacific PanDeist, Palpatine, Parkwells, Paulmcdonald, Pavel Vozenilek, Paxsimius, Pearl yam mirvana, Peasreach, Pekinpekin, Peli1414, Pepper, Persian Poet Gal, Peter Shearan, PeterSymonds, Petrb, Pgan002, Pharaoh of the Wizards, Phi beta, Phil Boswell, PhilKnight, Philip Trueman, Phkenyon, Pigkiller1127, Pinethicket, Pinksparky27, Platoon10, Plumbago, Podruznik, Postdlf, Prari, Prattflora, ProhibitOnions, Proyster, Ptmflbcs, Puchiko, Purplebackpack89, Qertis, Qqqqqq, Queson, Quidam65, Quinsareth, Quintote, Quirk, Qutezuce, RFD, RL0919, RaccoonFox, Ravenswing, Rayc, Rcawsey, Rdsmith4, Realm of Shadows, RedWolf, Reddi, Redraider1990, Renaissancee, Renato Caniatti, Revas, Rich Farmbrough, Rick Block, RickK, Ricky81682, Rjensen, Rjwilmsi, Rlquall, Romanm, Rompe, Root Beers, Rose Garden, Rreagan007, Rsg, Runnynose47, Rushadthomas, Ruskinmonkey, SNIyer12, SPUI, SWAdair, Salsb, Samisamij, Samuel Blanning, Saopaulo1, SatyrTN, Scanlan, Schnubble, Sciurinæ, Scott REDD, Scott5114, Scottman95, ScreaminEagle, Sdgjake, Seb az86556, Secret (renamed), Securiger, Sfmontyo, Shalom Yechiel, Shirik, Shpoon, Shreyu97, Shupchris, Sifaka, Simon J Kissane, Simon_J_Kissane, SirJello37, Sklauril, Skyhook41, Slambo, Slark, Slshow, Smarkflea, Smith03, Solipsist, Some jerk on the Internet, Someguy1221, Sophina1301, Spellcast, SpikeJones, Spydrlink, Srstoll37, Staffwaterboy, Stainskool, Stay cool, SteveStevic, Stwalkerster, Swid, Swid v.2, Swollib, TBadger, TUF-KAT, Tabletop, Tanaats, Tbhotch, Technogreek43, Teknomage, Teleomatic, Teles, Template namespace initialisation script, Tentontunic, Tetraedycal, Thaimoss, The Epopt, The High Fin Sperm Whale, The Interior, The Thing That Should Not Be, The Transhumanist, The Way, TheHoosierState89, TheOuthouseMouse, Thedjatclubrock, Thekohser, Thermal0xidizer, Thetomtom, Thingg, Tide rolls, Tidying Up, Timneu22, Timstevens890, Tinton5, TinyMAK5, Tiptoety, Titoxd, Tjss, Tm1000, Toby Bartels, TommyBoy, Tournesol, Tpbradbury, Tr-the-maniac, Tracer9999, Trimalchio, Triona, Triplechase, Trusilver, TurboV8, Turover, Twirlergirlkk, Tylertm, UPNSP, UkPaolo, Ulmanor, Uncle Dick, Unyoyega, Utcursch, Variousbagels, Vary, Velvetron, Vilcxjo, VinculumMan, Vipinhari, Viridian, Viriditas, Vishnava, Vivio Testarossa, Vrenator, Vsmith, Vzbs34, W3nwswiki, WAvegetarian, WOSlinker, WadeSimMiser, Waggers, Walor, Walton One, Wapcaplet, Wars, WarthogDemon, Waterguy, Weatherman78, Wereon, Wgbc2032, Whiskey in the Jar, WhisperToMe, WikiMaster12345, Wikipedian2, Wikipir8, Willking1979, Windchaser, Winterst, Wknight94, Wmahan, Woland37, Woohookitty, WormBobby, XLR2, XRK, Xiahou, Xymmax, Yamamoto Ichiro, Yayme23, Yockman, Youngamerican, YourNight, Yuckfoo, ZX81, Zeroxysm, Zoe, Zscout370, Île flottante, 1281 anonymous edits

Republican River *Source*: http://en.wikipedia.org/w/index.php?title=Republican_River *Contributors*: Ammodramus, Andres, AssistantX, Backspace, Bryan Derksen, Buaidh, CaroleHenson, DandyDan2007, Decumanus, FUBAR007, Footwarrior, Freechild, Hmains, Jllm06, Ken Gallager, Kswaterguys, Liblamb, Lightmouse, Paulmcdonald, Pfly, Smarkflea, Suntag, Tabletop, Twp, Wainstead, Woohookitty, Zeaner, 13 anonymous edits

Reservoir *Source*: http://en.wikipedia.org/w/index.php?title=Reservoir *Contributors*: Aatox, Acalamari, Alansohn, Allstarecho, AmericanLeMans, Amniarix, Andonic, Arjayay, Arthena, Astirmays, Awickert, B.d.mills, BL Lacertae, Babakathy, Bassbonerocks, Billinrio, Blehfu, Bobo192, Brougham96, Camw, Can't sleep, clown will eat me, CanadianLinuxUser, Camildo, Cassowary, Cctoide, Ceranthor, Chl, Chris j wood, ChrisCork, Ciphers, CommonsDelinker, Courcelles, Csmart287, CustardJack, Cyfal, DFS454, DVD R W, Davandron, Deror avi, DidiWeidmann, Dj Capricorn, Dnj710, Docu, Donarreiskoffer, Drunken Pirate, ENeville, ESkog, Em3rald, Emerson7, Eranjenes2, Erik9, Ewen, Ezhiki, FayssalF, Fieldday-sunday, Fifth Fish Finger, Floul1, Frehley, Fribbler, Frokor, Funandtrvl, Fæ, GLaDOS, Gail, GavinSharp, Gcm, GeoWriter, Geoff Bonanno, Geonarva, Gersenes, Ghirlandajo, Gilliam, HarlandQPitt, Hertz1888, HexaChord, Hike395, Hires an editor, Hmains, Ian n-s, Infrogmation, Ingolfson, J.Gifford, JBellis, JCDenton2052, JForget, Jacksieber, JamesAM, Jeffrey Mall, JetLover, Jim.henderson, Jonpollnow, Jose Icaza, Jpbowen, KVDP, Kaiba, Karlos87, Kbthompson, Kimse, Kjetil r, Kjkolb, Kostisl, Krj373, Kschwerdt514, Levineps, Lobob5, Logan, Lysy, Mailseth, Makeemlighter, Malathos, Marcia Wright, Martial75, Mattgirling, Mattisse, MaxSem, MeekMark, Metropolitan90, MichaelMaggs, Minghong, Mircea cs, Moanzhu, Moreau1, Muijz, Muéro, Narson, Newell Post, Nigholith, Nihiltres, Nixsun01, Nk, NortyNort, Nshimbi, Ohdear15, Oismiffy, OlEnglish, Old Moonraker, Orphan Wiki, Osarius, Pengo, Pete ess, Peter gk, Peterlewis, Philip Trueman, Pigsonthewing, Planetjanet, Pointillist, PopUpPirate, Prabinepali, Prashanthns, Qfl247, RJFJR, Radiojon, Rajasekhar1961, Ravedave, Redmegtheavenger, Rich Farmbrough, Robina Fox, Roxi2, Rruss, S4n1HS22WMX695In, Samnz05, SchuminWeb, Sean the Spook, Severnapark, Sfahey, Shannon1, Sole Soul, Squids and Chips, Svetovid, Swanml, Tabletop, Tedickey, TenOfAllTrades, Tgeairn, TheGerm, Thingg, Thw1309, Tide rolls, Tills, Tom harrison, Totse-vandal, Tranquil demon, Trevyn, Trilby*foxglove, Tsuite, Túrelio, Uncle Dick, Unschool, VPliousnine, Vdegroot, Vegaswikian, Velella, Versus22, Vhorvat, VinceBowdren, Wafulz, Walrusfunk, Wavelength, Whosasking, Wikid77, Wikijmms, William Avery, YixilTesiphon, 250 anonymous edits

Phillips County, Kansas *Source*: http://en.wikipedia.org/w/index.php?title=Phillips_County%2C_Kansas *Contributors*: Afruit, Backspace, Bel89, CanisRufus, CommonsDelinker, D6, DavidSteinle, Grumpykansan, Hephaestos, Leslie Mateus, Moverton, Nyttend, Omnedon, RSM, Ram-Man, Rich Farmbrough, Rjwilmsi, Sbmeirow, Sgt Pinback, Smallbones, TUF-KAT, Template namespace initialisation script, Theduckman1763, UtherSRG, Whhalbert, WillC, Worldenc, 5 anonymous edits

Colorado *Source*: http://en.wikipedia.org/w/index.php?title=Colorado *Contributors*: 0, 1-800-94-Jenny, 16@r, 1Martin33, 1brettsnyder, 23skidoo, 40Degrees, 63.215.230.xxx, 7mike5000, A F K When Needed, ALT2870, ARUenergy, Aaron, Abog, Acalamari, Achangeisasgoodasa, Acroterion, Adagio, Adamciks, Adrian.benko, Aesopos, After Midnight, Agassiz830, AgentPeppermint, Ahassan05, Ahoerstemeier, Aitias, Aktron, Alansohn, AlexiusHoratius, Alexmcfire, Alexwcovington, Allstarecho, Altairisfar, Alvaro, Amazonien, AmeriCan, Anandks007, Ancheta Wis, Andy Marchbanks, Andy120290, Angela, Angr, Anonymous Dissident, Anonymous editor, Antandrus, AntsGoMarching123, Anturiaethwr, Arakunem, Arjun01, Astroview120mm, Astuishin, Atanamir, AtticusX, AuburnPilot, AuburnPilot, Aude, Austin72900, Avaragado, Aztom2, BGManofID, BRG, BSveen, Babajobu, BabuBhatt, Ballsinyourface, Barbra123, Barek, Barneca, Baseball 292992, Beezer137, Bejinhan, Beland, Bender235, Benjh40, Bhumiya, Bigh, Bjwebb, Bkonrad, BlaiseFEgan, BlueAzure, Blueboy96, Bluedogtn, Bob790, Bobianite, Bobjoe3, Bobjohn69, Bobo192, Bomac, Bongwarrior, BookWormHR, BoomMan77, BorgHunter, Bped1985, Breast Implants, Brewmanz, Brian the Editor, Bsod2, Buaidh, Bucketsofg, Buddha24, Burnwelk, Butwhatdoiknow, C5813, CASportsFan, CW221, Cabhan, Cablebfg, Caesura, Caiaffa, Caltas, CambridgeBayWeather, Camw, Cananadian402, Canglesea, Canthusus, Capecodeph, Capnsuite, Caponer, Capricorn42, CaptainJae, Carlaude, Carptrash, Carter98, Cburnett, Cchow2, Celestra, Censusdata, Ceranthor, Ceyockey, Cflores77, Chadlupkes, Charolastra charolo, Chevelle-Marie, Chris Roy, Chris the speller, ChrisRuvolo, Chrishmt0423, Chun-hian, Chzz, Ciphers, Civil Engineer III, Ckatz, Ckimpson, ClairSamoht, ClockworkSoul, Closedmouth, Cmaestas1, Cmr08, Code36, Coelacan, ColoradoZ, Cometstyles, CompRhetoric, Conversion script, CopperSquare, Corpx, Coulraphobic123, Cowman141414, Cp111, Cranky305, CrazyC83, Cribbswh, Crobzub, Cronium, Curty Wurty, CutOffTies, Cyrius, DARTH SIDIOUS 2, DCEdwards1966, DHeyward, DVdm, DXRAW, Dabomb87, Dallas Hays, Damien solta, DanielCD, Danny, Darrel M, Dave6, Davehi1, Daven200520, David Wahler, David.Monniaux, Db099221, Dbenbenn, Dcornwall, Deadlyops, Debresser, Decumanus, Deenoe, Deepshadows88, Deflective, Delldot, DenverDude99, Denverjason, Deor, DerHexer, Deucesmd, DevDev22, Devin Pederson, Dharh, Diannaa, Dino, Discospinster, Discoverydude, DividedByNegativeZero, DivineIntervention, Dkdantastic, Dlohcierekim's sock, Docu, Dolovis, Dominic.klaassen, Doovinator, DoubleBlue, Doublenickle59, Dr. Blofeld, Dr4chess, Dream-box777, Dreist, Droll, Drumlineramos, Dtpeck, Durang Dorad, Dwillsey, Dwilso, Dwo, Dylan Lake, DynamoDegsy, Dysprosia, E2eamon, ESkog, Eagle4000, Eco84, Ed g2s, Edderso, Eddy15467, Editor19841,

EditorASC, Edivorce, Edsully, Edwardadrian, Eiceman, Eitoki, Ejdjjd, Elchampion, Eliz81, Ellywa, Eloquence, Emeraldcityserendipity, Emkind2, Emote, EncMstr, EncycloPetey, Enkrates, Enviroboy, Epbr123, Epicv27, Epolk, Eragonkid10, ErgoSum88, Eric outdoors, Eric-Wester, Erzahler, Escape Orbit, Eu.stefan, Evice, Extreme outdoors, Exxolon, Fabalous Sex, Farewellbitch, Fastilysock, Fayenatic london, Feliz, Fg2, Finalnight, Fishing, Flatterworld, Flewis, Flipper89799, Fluffernutter, Fnlayson, Fogherty V. Tatin, Footwarrior, Forever Dusk, Fratrep, Fred Bauder, Fredbauder, Frietjes, FrozenPurpleCube, Funnyhat, Fusionmix, FyzixFighter, GL, Gabe Radovsky, Gaiacarra, Gaius Cornelius, Galoubet, Gator1, Gbalitgs, Gellersen, Gene Nygaard, Geni, Geo.plrd, Geoffrey, Geologyguy, Gerrish, Ghettodub, Giants27, Ginkgo100, GirasoleDE, Glenn666, GlobeTrotter, Gnative, Gnoffprince, Gobeirne, Goethean, Golbez, Good Olfactory, Gpietsch, Grace1019778, GraemeMcRae, Graham87, Grand51paul, GrantG9, Grawpy, Greepnik, GregU, Gsarwa, Gugilymugily, Guliolopez, Gump Stump, GunFactor007, Gwk, H0rsl0ver, HALLObøøøøh, Hadal, Hahathatswrong, Halcionne, Hammersoft, Harburg, Harkamal1001, Hasek is the best, Hawk4free, Heegoop, Heptapod, His Manliness, Hmains, Hobartimus, Hogs555, Hondje, Hopiakuta, Horologium, Hp69boy, Hu12, Husond, Hydrogen Iodide, Hydrox, I dont know is it, Iamapuppy, Ianthegecko, Iantheinvincible, Icek, Icepick09, Ihoney, Ilovechicken321, Imanfan, Imnotminkus, ImperatorExercitus, Includejapan, Inka 888, Inliten, Iridescent, Ishmaell, J Di, J.delanoy, JForget, JFreeman, JLaTondre, JNW, JSpung, JW1805, Jack Cox, Jackerstine, Jacob.jose, Jasmyn131313, Jauhienij, Jaycasper, Jcam, Jdforrester, Jean Xu, Jeff G., Jengod, Jeremiestrother, Jerome709, Jeronimo, Jhendin, Jim Horning, JimIrwin, Jimeree, JinJian, Jkominek, Jlkventures, Jmdickman, Jmlk17, Jodyreale, JoeSmack, Joetorre, John, John K, JohnInDC, Jojhutton, Jonstjohn, Jorunn, Jose77, Jsding, Juantay, Jude1955, Judeeclare, Juliancolton, Jultemplet, Justin Eiler, Justinstroud, Juzeris, Jwick, K. Annoyomous, KF, Kablammo, Kahuroa, Kar54, Katalaveno, Kazrak, Kbh3rd, Keith D, Keith Edkins, KeithB, Kellybean4, Ken g6, Keraunoscopia, Keylay31, Khaenman11, King of Hearts, Kintetsubuffalo, Kkinder, Klatoo55, Kmanblue, Kmccabe2003, Knadaves, KnowledgeOfSelf, Koavf, Kontar, Kornfan71, Kostisl, Koyaanis Qatsi, Kozuch, Krawi, Krellis, Kristinpedia, Kusma, Kwamikagami, LDHan, LOL, LabradorLover456, Lamb Kebab, Lan56, LanceBarber, Laubacht, Laurenrose3091, Leahcim506, Levineps, Lightdarkness, Lightmouse, Lights, Lilac Soul, LinDrug, Lincolnite, Ling.Nut, LinkTiger, Linnix, LizardJr8, Loco830, Logan, Longnskinny, Lorax, Lord Voldemort, Loren36, Lotje, Luke4545, Luna Santin, Luong, M C Y 1008, MER-C, MJCdetroit, MONGO, MPF, MacsBug, Maddie!, Maelnuneb, Magioladitis, Magumbaboy, Manop, Manuel Trujillo Berges, Maramanc, MarkSutton, Marnanel, Materialscientist, Matijap, MattWright, Mausy5043, Mav, Mboverload, McCaster, Mcecil82, Mcpusc, Melaen, Mephistophelian, Michael Hardy, Michael Patrick, Midnightcomm, Mike Rosoft, Mike s, MikeLynch, Mikegreve, Mikevegas40, Miles2244, Miles530, Minesweeper, Minesweeper.007, Mintchocolatebear, Mintleaf, Miranda, Mirv, Miskwito, Mlaffs, Molas, Momo san, Mononomic, Monterey Bay, Monty845, Morwen, Moverton, MrRadioGuy, MrsDeMent, Mu Cow, Mwanner, Mxn, MyColoradoNightlife, NE2, NEDIB AMABO, NEICenergy, NJPharris, Nakon, NameIsRon, Nasnema, Natalie Erin, NawlinWiki, Nehrams2020, Nepenthes, NerdyScienceDude, Neutrality, Never give in, NewGuy4, Ngfan1, Nihiltres, Nikkita45678, Nikkiw1996, NinjaCharlie, Nixeagle, Nkrosse, Nlu, Northwesterner1, Notchcode, Num1scot, Numbo3, Nyyg212, Odelta2011, Odie5533, Ohnoitsjamie, Ohwilleke, Oman66, One+1=two, Onetwo1, Orang gila, Orange Suede Sofa, Originalwana, Orlady, Otets, Outburn, OverSS, Owen, OwenBlacker, Oxymoron83, PASTAPUSS, PDH, PITTPANTS121, Packerfansam, Panchitaville, Paris1127, Patrick, Patrickesmonde, Pb30, Pckrfan375, Pekinpekin, Perfect Proposal, Persian Poet Gal, Peter Winnberg, PeterSymonds, Pharaoh of the Wizards, Philip Trueman, PhilipO, Philosopher, Phyllis1753, Piano non troppo, PianoMelody, Picabo, Pigman, Pinethicket, Pingpongmaster09, Pjc2025, Plasticup, Plazak, Podruznik, Postdlf, Pras, Prattflora, Prestonmag, Prof. Spaghetti, PuppyAddict, Pupster21, Pxma, Pyrospirit, Qqqqqq, Quadell, QueenofBattle, Queson, R.T.Gellar, RL0919, RMHED, RSM, Radiojon, Radon210, RainbowOfLight, Ramirez72, Rathersane, Ravedave, Ravenmasterq, Ravensfan5252, Ray Van De Walker, Rdivins, Realkyhick, Realm of Shadows, Red knots, RedMachine99, Reddi, Reesh, Regancy42, Renato Caniatti, Rettetast, Revas, RexNL, Rfc1394, Rfl, Rhatsa26X, Rhoades320, Rhobite, Rich Farmbrough, Richard D. LeCour, Rick Block, Ritcheyrp, RjLesch, Rjwilmsi, Rlquall, Rmcfan12, Rmhermen, RobLa, Robertb-dc, Rock616, Rocketmaniac, Rodrigouf, Rodyx, Rohit147, Romanm, Root Beers, Rossdegenstein, Roy Brumback, RoyBoy, Rrburke, Rreagan007, Rror, Ruhrfisch, Runningonbrains, Rupertslander, RyanMK83, S11080, S4n1HS22WMX695In, SCEhardt, SJP, SNIyer12, SPUI, ST47, Sam Blacketer, Samscarb, Samuel Blanning, Sanfranman59, Sango123, SatyrTN, Scanlan, Scapler, Scarian, Schiptuin, Schneelocke, Scott5114, Scriberius, Seanth123, Secret (renamed), Sedna10387, Seidenstud, SelGomez, Seraphim, Seth Ilys, Severinus, Sfan00 IMG, Sfmontyo, Shalom Yechiel, SheepNotGoats, Shimgray, Shimmera, Shinglor, ShiningPath, Shpoon, Sift&Winnow, Silverfish123, SimMoonXP, Simoneau, Sjakkalle, Skarebo, Skier Dude, Slowking Man, Smalljim, Smallman12q, Smith03, Snigbrook, Snowolf, SoulMeetsBody, South Bay, Spain21, Spincycle, SpuriousQ, Spydrlink, Staffwaterboy, Staplegunther, Steel, Stephen Bain, Stepp-Wulf, Stevertigo, Stevewonder2, Strongbad1982, Struhs, SunDragon34, Sunray, Supercon724, Svgalbertian, Svick, Swid, SwisterTwister, Swollib, Sylvain1972, THEN WHO WAS PHONE?, TJRC, TUF-KAT, Tanthalas39, Tarfu92, Tarkan1st, Tatrgel, Tbhotch, Tedder, Tedickey, Tele editor, Template namespace initialisation script, Tesscass, Tgeairn, ThatDamnDave, Thatguyflint, The Epopt, The Obento Musubi, The Random Editor, The Thing That Should Not Be, The Transhumanist, The Way, The wub, TheCarterVI, TheDJ, TheRanger, Thepope0987, Thermal0xidizer, Thiseye, Thor21, Thrane, Tide rolls, Tidying Up, Tigermania78, Tim Ross, Tim1357, Timeineurope, Timneu22, Titoxd, Tobby72, Todd Vierling, Tom Danson, Tom harrison, Tommylane1, Tony1, TonyTheTiger, Tpbradbury, Tracer9999, Trevor MacInnis, Triwbe, Trodaikid1983, Tstockma, Tuan Dang, TwinCityIL, Uncle Milty, Unknown One, Unschool, Urbanus, Urok, Utcursch, Ute in DC, UtherSRG, Vailskier, Vanished User 8a9b4725f8376, Vanished user 39948282, Velella, Vertigo700, ViciousJ, Viper55, Viriditas, Vishnava, Vpuliva, Vsion, Vsmith, WadeSimMiser, Wagonsoul773, Wangi, Wapcaplet, Wars, Waterjuice, Watkinsian, Way wiki, Wayne Slam, Weathermaps, Webdinger, Webster6, Wester, Whisky drinker, WhisperToMe, Wik, Wiki alf, Wikipelli, Will Beback, William Allen Simpson, Willking1979, Wimt, Wizardman, Wknight94, Wmahan, Woohookitty, Work permit, XRK, XkillzoneX, Xnatedawgx, Xodl2x1o, Xyzzyva, Yamamoto Ichiro, Yarnalgo, YellowMonkey, Youngamerican, Zephyr2k, Zoddette, Zodon, Zoe, Zollerriia, Zro, Zscout370, Zstroup22, Zundark, Zzyzx11, 1662 anonymous edits

Republican City, Nebraska *Source*: http://en.wikipedia.org/w/index.php?title=Republican_City%2C_Nebraska *Contributors*: Bumm13, Cricketcatcher, DandyDan2007, Freechild, Jllm06, John Cardinal, Nyttend, Ram-Man, 5 anonymous edits

Alma, Nebraska *Source*: http://en.wikipedia.org/w/index.php?title=Alma%2C_Nebraska *Contributors*: Accurizer, Acntx, Ammodramus, Bumm13, CheepnisAroma, Cricketcatcher, Finngall, Freechild, Hajatvrc, Hmains, Iridescent, Ja 62, John Cardinal, Ks0stm, Lightmouse, Nyttend, Pearle, Qharouff, RWReagan, Ram-Man, Rich Farmbrough, ScottyBerg, Semperf, Stepp-Wulf, Swid, Tibetan Prayer, Tonythetiger6, Zdv, 31 anonymous edits

Brownville, Nebraska *Source*: http://en.wikipedia.org/w/index.php?title=Brownville%2C_Nebraska *Contributors*: Americasroof, Ammodramus, Bumm13, DandyDan2007, Freechild, Ground Zero, Hmains, Jllm06, John Cardinal, Korg, MrBill47, Nyttend, Onyourside, Pearle, Pink!Teen, Ram-Man, Rich Farmbrough, T. Anthony, Tjc, Zigzig20s, Zurmablue, 6 anonymous edits

Image Sources, Licenses and Contributors

Image:Republican-river.png *Source*: http://en.wikipedia.org/w/index.php?title=File:Republican-river.png *License*: unknown *Contributors*: Original uploader was Liblamb at en.wikipedia

File:Flag of Nebraska.svg *Source*: http://en.wikipedia.org/w/index.php?title=File:Flag_of_Nebraska.svg *License*: unknown *Contributors*: Ammodramus, Anime Addict AA, Denelson83, Dzordzm, Fry1989, Homo lupus, Serinde, Svgalbertian, Zscout370, 2 anonymous edits

File:Nebraska-StateSeal.svg *Source*: http://en.wikipedia.org/w/index.php?title=File:Nebraska-StateSeal.svg *License*: unknown *Contributors*: U.S. Government

File:Nebraska in United States.svg *Source*: http://en.wikipedia.org/w/index.php?title=File:Nebraska_in_United_States.svg *License*: unknown *Contributors*: TUBS

Image:Speakerlink.svg *Source*: http://en.wikipedia.org/w/index.php?title=File:Speakerlink.svg *License*: unknown *Contributors*: Woodstone. Original uploader was Woodstone at en.wikipedia

Image:Nebraska 1718.jpg *Source*: http://en.wikipedia.org/w/index.php?title=File:Nebraska_1718.jpg *License*: unknown *Contributors*: Original uploader was Billwhittaker at en.wikipedia

Image:Scotts bluff national monument.jpg *Source*: http://en.wikipedia.org/w/index.php?title=File:Scotts_bluff_national_monument.jpg *License*: unknown *Contributors*: User:Podruznik

Image:Nebraska.JPG *Source*: http://en.wikipedia.org/w/index.php?title=File:Nebraska.JPG *License*: unknown *Contributors*: User:ErgoSum88

Image:National-atlas-nebraska.PNG *Source*: http://en.wikipedia.org/w/index.php?title=File:National-atlas-nebraska.PNG *License*: unknown *Contributors*: Abhijitsathe

Image:Nebraska population map.png *Source*: http://en.wikipedia.org/w/index.php?title=File:Nebraska_population_map.png *License*: unknown *Contributors*: JimIrwin at en.wikipedia

File:I-76.svg *Source*: http://en.wikipedia.org/w/index.php?title=File:I-76.svg *License*: unknown *Contributors*: Augiasstallputzer, Fran Rogers, Ltljltlj, SPUI, 1 anonymous edits

File:I-80.svg *Source*: http://en.wikipedia.org/w/index.php?title=File:I-80.svg *License*: unknown *Contributors*: User:Ltljltlj

File:I-129.svg *Source*: http://en.wikipedia.org/w/index.php?title=File:I-129.svg *License*: unknown *Contributors*: Kanonkas, Ltljltlj, SPUI, Xnatedawgx

File:I-180.svg *Source*: http://en.wikipedia.org/w/index.php?title=File:I-180.svg *License*: unknown *Contributors*: Ltljltlj, Rocket000, SPUI, Xnatedawgx

File:I-480.svg *Source*: http://en.wikipedia.org/w/index.php?title=File:I-480.svg *License*: unknown *Contributors*: Ltljltlj, SPUI, Xnatedawgx, 1 anonymous edits

File:I-680.svg *Source*: http://en.wikipedia.org/w/index.php?title=File:I-680.svg *License*: unknown *Contributors*: Ltljltlj, SPUI, Xnatedawgx, 1 anonymous edits

File:US 6.svg *Source*: http://en.wikipedia.org/w/index.php?title=File:US_6.svg *License*: unknown *Contributors*: Lpangelrob, Rocket000, SPUI

File:US 20.svg *Source*: http://en.wikipedia.org/w/index.php?title=File:US_20.svg *License*: unknown *Contributors*: Kanonkas, SPUI

File:US 26.svg *Source*: http://en.wikipedia.org/w/index.php?title=File:US_26.svg *License*: unknown *Contributors*: Bidgee, SPUI, 3 anonymous edits

File:US 30.svg *Source*: http://en.wikipedia.org/w/index.php?title=File:US_30.svg *License*: unknown *Contributors*: Bidgee, SPUI, 4 anonymous edits

File:US 34.svg *Source*: http://en.wikipedia.org/w/index.php?title=File:US_34.svg *License*: unknown *Contributors*: Bidgee, SPUI, Xnatedawgx, 3 anonymous edits

File:US 73.svg *Source*: http://en.wikipedia.org/w/index.php?title=File:US_73.svg *License*: unknown *Contributors*: Bidgee, SPUI, 2 anonymous edits

File:US 75.svg *Source*: http://en.wikipedia.org/w/index.php?title=File:US_75.svg *License*: unknown *Contributors*: SPUI, Xnatedawgx, 1 anonymous edits

File:US 77.svg *Source*: http://en.wikipedia.org/w/index.php?title=File:US_77.svg *License*: unknown *Contributors*: Bidgee, SPUI, Xnatedawgx, 2 anonymous edits

File:US 81.svg *Source*: http://en.wikipedia.org/w/index.php?title=File:US_81.svg *License*: unknown *Contributors*: Bidgee, Hellbus, SPUI, Xnatedawgx, 2 anonymous edits

File:US 83.svg *Source*: http://en.wikipedia.org/w/index.php?title=File:US_83.svg *License*: unknown *Contributors*: Bidgee, SPUI, Xnatedawgx, 2 anonymous edits

File:US 136.svg *Source*: http://en.wikipedia.org/w/index.php?title=File:US_136.svg *License*: unknown *Contributors*: Einstein2, SPUI

File:US 138.svg *Source*: http://en.wikipedia.org/w/index.php?title=File:US_138.svg *License*: unknown *Contributors*: User:Vishwin60

File:US 159.svg *Source*: http://en.wikipedia.org/w/index.php?title=File:US_159.svg *License*: unknown *Contributors*: SPUI

File:US 183.svg *Source*: http://en.wikipedia.org/w/index.php?title=File:US_183.svg *License*: unknown *Contributors*: SPUI, 2 anonymous edits

File:US 275.svg *Source*: http://en.wikipedia.org/w/index.php?title=File:US_275.svg *License*: unknown *Contributors*: SPUI

File:US 281.svg *Source*: http://en.wikipedia.org/w/index.php?title=File:US_281.svg *License*: unknown *Contributors*: SPUI, Xnatedawgx

File:US 283.svg *Source*: http://en.wikipedia.org/w/index.php?title=File:US_283.svg *License*: unknown *Contributors*: SPUI

File:US 385.svg *Source*: http://en.wikipedia.org/w/index.php?title=File:US_385.svg *License*: unknown *Contributors*: SPUI, Xnatedawgx, 1 anonymous edits

image:Nebraska quarter, reverse side, 2006.jpg *Source*: http://en.wikipedia.org/w/index.php?title=File:Nebraska_quarter,_reverse_side,_2006.jpg *License*: unknown *Contributors*: Dbenbenn, H-stt, 2 anonymous edits

File:Nebraska State Capitol at night.jpg *Source*: http://en.wikipedia.org/w/index.php?title=File:Nebraska_State_Capitol_at_night.jpg *License*: unknown *Contributors*: Decumanus at en.wikipedia

Image:Downtown Omaha Skyline 2010.jpg *Source*: http://en.wikipedia.org/w/index.php?title=File:Downtown_Omaha_Skyline_2010.jpg *License*: unknown *Contributors*: JonClee86

Image:Lincoln ne skyline.JPG *Source*: http://en.wikipedia.org/w/index.php?title=File:Lincoln_ne_skyline.JPG *License*: unknown *Contributors*: Admrboltz, Ebyabe, EugeneZelenko, Malepheasant, Stack

Image: RepublicanRiver1947.jpg *Source*: http://en.wikipedia.org/w/index.php?title=File:RepublicanRiver1947.jpg *License*: unknown *Contributors*: J.G. Connor (submitted to USGS by Steve Blanchard, OSW) Original uploader was Suntag at en.wikipedia

Image: Republican-river.png *Source*: http://en.wikipedia.org/w/index.php?title=File:Republican-river.png *License*: unknown *Contributors*: Original uploader was Liblamb at en.wikipedia

Image:RepublicanRiverBasin.png *Source*: http://en.wikipedia.org/w/index.php?title=File:RepublicanRiverBasin.png *License*: unknown *Contributors*: Kansas Department of Agriculture

Image:Taiwan JungHua Dam.JPG *Source*: http://en.wikipedia.org/w/index.php?title=File:Taiwan_JungHua_Dam.JPG *License*: unknown *Contributors*: User:Vegafish

Image:Lakevyrnwysummer.jpg *Source*: http://en.wikipedia.org/w/index.php?title=File:Lakevyrnwysummer.jpg *License*: unknown *Contributors*: User:Sean the Spook. Original uploader was Sean the Spook at en.wikipedia

Image:Stocks Reservoir.jpg *Source*: http://en.wikipedia.org/w/index.php?title=File:Stocks_Reservoir.jpg *License*: unknown *Contributors*: User:Karlos87

Image:GibsonR.jpg *Source*: http://en.wikipedia.org/w/index.php?title=File:GibsonR.jpg *License*: unknown *Contributors*: User:Citypeek

Image:Hydroelectric dam.svg *Source*: http://en.wikipedia.org/w/index.php?title=File:Hydroelectric_dam.svg *License*: unknown *Contributors*: User:Tomia

Image:KupferbachStauseeAachen.jpg *Source*: http://en.wikipedia.org/w/index.php?title=File:KupferbachStauseeAachen.jpg *License*: unknown *Contributors*: User:Túrelio

Image:Llyn Brianne spillway.jpg *Source*: http://en.wikipedia.org/w/index.php?title=File:Llyn_Brianne_spillway.jpg *License*: unknown *Contributors*: User:Velela

Image:volta lake.jpg *Source*: http://en.wikipedia.org/w/index.php?title=File:Volta_lake.jpg *License*: unknown *Contributors*: NASA

Image:Lake Kariba.jpg *Source*: http://en.wikipedia.org/w/index.php?title=File:Lake_Kariba.jpg *License*: unknown *Contributors*: Martin H., Mehmet Karatay, Paobac, TheDJ, TommyBee, 1 anonymous edits

File:Phillips Co KS Courthouse.JPG *Source*: http://en.wikipedia.org/w/index.php?title=File:Phillips_Co_KS_Courthouse.JPG *License*: unknown *Contributors*: User:Smallbones

File:Map of Kansas highlighting Phillips County.svg *Source*: http://en.wikipedia.org/w/index.php?title=File:Map_of_Kansas_highlighting_Phillips_County.svg *License*: unknown *Contributors*: User:Dbenbenn

File:Map of USA KS.svg *Source*: http://en.wikipedia.org/w/index.php?title=File:Map_of_USA_KS.svg *License*: unknown *Contributors*: Abnormaal, EurekaLott, Huebi, Lokal Profil, Lupo, Mattbuck, Petr Dlouhý, 2 anonymous edits

Image:USA Phillips County, Kansas age pyramid.svg *Source*: http://en.wikipedia.org/w/index.php?title=File:USA_Phillips_County,_Kansas_age_pyramid.svg *License*: unknown *Contributors*: user:WarX

Image:Map of Phillips Co, Ks, USA.png *Source*: http://en.wikipedia.org/w/index.php?title=File:Map_of_Phillips_Co,_Ks,_USA.png *License*: unknown *Contributors*: Designed and published by the Kansas Department of Transportation.

File:Flag of Colorado.svg *Source*: http://en.wikipedia.org/w/index.php?title=File:Flag_of_Colorado.svg *License*: unknown *Contributors*: User:Fleminra

File:Seal of Colorado.svg *Source*: http://en.wikipedia.org/w/index.php?title=File:Seal_of_Colorado.svg *License*: unknown *Contributors*: User:Svgalbertian

File:Colorado in United States.svg *Source*: http://en.wikipedia.org/w/index.php?title=File:Colorado_in_United_States.svg *License*: unknown *Contributors*: TUBS

File:Tenmile Fletcher Group.jpg *Source*: http://en.wikipedia.org/w/index.php?title=File:Tenmile_Fletcher_Group.jpg *License*: unknown *Contributors*: Wayne L. Bart. Original uploader was Throughthelens at en.wikipedia

File:6601-RedRocksTheater.JPG *Source*: http://en.wikipedia.org/w/index.php?title=File:6601-RedRocksTheater.JPG *License*: unknown *Contributors*: Robert J. Boser, EditorASC (talk) 10:12, 24 November 2008 (UTC). Original uploader was EditorASC at en.wikipedia

File:ChicoClosedBasin.jpg *Source*: http://en.wikipedia.org/w/index.php?title=File:ChicoClosedBasin.jpg *License*: unknown *Contributors*: User:Fred Bauder

File:LovelandPass CMM.jpg *Source*: http://en.wikipedia.org/w/index.php?title=File:LovelandPass_CMM.jpg *License*: unknown *Contributors*: Christopher Morrison

File:Mount Elbert2.JPG *Source*: http://en.wikipedia.org/w/index.php?title=File:Mount_Elbert2.JPG *License*: unknown *Contributors*: Hogs555

File:6408-LovelandPassCOLO.jpg *Source*: http://en.wikipedia.org/w/index.php?title=File:6408-LovelandPassCOLO.jpg *License*: unknown *Contributors*: User:EditorASC

File:Grand Mesa and Mt. Garfield.jpg *Source*: http://en.wikipedia.org/w/index.php?title=File:Grand_Mesa_and_Mt._Garfield.jpg *License*: unknown *Contributors*: User:Tewy

File:DNM Morrison.jpg *Source*: http://en.wikipedia.org/w/index.php?title=File:DNM_Morrison.jpg *License*: unknown *Contributors*: Anky-man 16:26, 16 April 2007 (UTC). Original uploader was Anky-man at en.wikipedia

File:Bearlakeinspring2.jpg *Source*: http://en.wikipedia.org/w/index.php?title=File:Bearlakeinspring2.jpg *License*: unknown *Contributors*: User:Charlessauer

File:PivotWithDrops.JPG *Source*: http://en.wikipedia.org/w/index.php?title=File:PivotWithDrops.JPG *License*: unknown *Contributors*: H2O-C, Honeplus, NJR ZA, 2 anonymous edits

File:GrandJunctionTrip92007013.jpg *Source*: http://en.wikipedia.org/w/index.php?title=File:GrandJunctionTrip92007013.jpg *License*: unknown *Contributors*: User:Staplegunther

File:October Snow in Colorado.jpg *Source*: http://en.wikipedia.org/w/index.php?title=File:October_Snow_in_Colorado.jpg *License*: unknown *Contributors*: Jeff Schmaltz

File:Mesa-Verde---Cliff-Palace-in 1891 - edit1.jpg *Source*: http://en.wikipedia.org/w/index.php?title=File:Mesa-Verde---Cliff-Palace-in_1891_-_edit1.jpg *License*: unknown *Contributors*: User:Andrew c

File:Beol court25.jpg *Source*: http://en.wikipedia.org/w/index.php?title=File:Beol_court25.jpg *License*: unknown *Contributors*: unknown NPS photographer

File:Wpdms kansas nebraska utah territories 1860 idx.png *Source*: http://en.wikipedia.org/w/index.php?title=File:Wpdms_kansas_nebraska_utah_territories_1860_idx.png *License*: unknown *Contributors*: Wars, 1 anonymous edits

File:Denver 1859.jpg *Source*: http://en.wikipedia.org/w/index.php?title=File:Denver_1859.jpg *License*: unknown *Contributors*: V14256 U.S. Copyright Office. Associated name on shelflist card: Collier & Cleveland Litho Co.

File:The-Battle-of-Glorieta-Pass.jpg *Source*: http://en.wikipedia.org/w/index.php?title=File:The-Battle-of-Glorieta-Pass.jpg *License*: unknown *Contributors*: unknown

File:Mount of the Holy Cross.jpeg *Source*: http://en.wikipedia.org/w/index.php?title=File:Mount_of_the_Holy_Cross.jpeg *License*: unknown *Contributors*: Dino, Editor19841, Monkeybait, Plazak, RedWolf

File:Georgetown loop 1899.jpg *Source*: http://en.wikipedia.org/w/index.php?title=File:Georgetown_loop_1899.jpg *License*: unknown *Contributors*: Infrogmation, Quatro Valvole, Trialsanderrors

File:Colorado population map.png *Source*: http://en.wikipedia.org/w/index.php?title=File:Colorado_population_map.png *License*: unknown *Contributors*: Original uploader was JimIrwin at en.wikipedia

File:Colorado Rocky Mtn Church.jpg *Source*: http://en.wikipedia.org/w/index.php?title=File:Colorado_Rocky_Mtn_Church.jpg *License*: unknown *Contributors*: User:Changlc

File:USAFA Chapel from terrazzo.JPG *Source*: http://en.wikipedia.org/w/index.php?title=File:USAFA_Chapel_from_terrazzo.JPG *License*: unknown *Contributors*: User:EpicV27

File:ColoradoHistoryMuseum.jpg *Source*: http://en.wikipedia.org/w/index.php?title=File:ColoradoHistoryMuseum.jpg *License*: unknown *Contributors*: User:Xnatedawgx

File:Denver Colorado Art.jpg *Source*: http://en.wikipedia.org/w/index.php?title=File:Denver_Colorado_Art.jpg *License*: unknown *Contributors*: User:7mike5000

File:Denver CO DT WTC AMH 322.JPG *Source*: http://en.wikipedia.org/w/index.php?title=File:Denver_CO_DT_WTC_AMH_322.JPG *License*: unknown *Contributors*: User:Wars

File:2006 CO Proof.png *Source*: http://en.wikipedia.org/w/index.php?title=File:2006_CO_Proof.png *License*: unknown *Contributors*: United States Mint

File:Corn production in Colorado.jpg *Source*: http://en.wikipedia.org/w/index.php?title=File:Corn_production_in_Colorado.jpg *License*: unknown *Contributors*: Severino666, Urban, 1 anonymous edits

File:DSCN2549 beevesnearwalden 600.jpg *Source*: http://en.wikipedia.org/w/index.php?title=File:DSCN2549_beevesnearwalden_600.jpg *License*: unknown *Contributors*: User:Decumanus

File:Grand Junction Trip 92007 131.JPG *Source*: http://en.wikipedia.org/w/index.php?title=File:Grand_Junction_Trip_92007_131.JPG *License*: unknown *Contributors*: Kintetsubuffalo, Staplegunther, 1 anonymous edits

File:Colorado.JPG *Source*: http://en.wikipedia.org/w/index.php?title=File:Colorado.JPG *License*: unknown *Contributors*: User:ErgoSum88

File:I-25 (CO).svg *Source*: http://en.wikipedia.org/w/index.php?title=File:I-25_(CO).svg *License*: unknown *Contributors*: Ltljltlj, T2, Xnatedawgx

File:US 84.svg *Source*: http://en.wikipedia.org/w/index.php?title=File:US_84.svg *License*: unknown *Contributors*: Bidgee, SPUI, Xnatedawgx, 2 anonymous edits

File:US 287.svg *Source*: http://en.wikipedia.org/w/index.php?title=File:US_287.svg *License*: unknown *Contributors*: SPUI, Xnatedawgx, 3 anonymous edits

File:I-70 (CO).svg *Source*: http://en.wikipedia.org/w/index.php?title=File:I-70_(CO).svg *License*: unknown *Contributors*: Ltljltlj, Nilfanion, T2, 2 anonymous edits

File:US 24.svg *Source*: http://en.wikipedia.org/w/index.php?title=File:US_24.svg *License*: unknown *Contributors*: Bidgee, SPUI, 3 anonymous edits

File:US 85.svg *Source*: http://en.wikipedia.org/w/index.php?title=File:US_85.svg *License*: unknown *Contributors*: SPUI, Xnatedawgx, 1 anonymous edits

File:US 350.svg *Source*: http://en.wikipedia.org/w/index.php?title=File:US_350.svg *License*: unknown *Contributors*: SPUI

File:I-76 (CO).svg *Source*: http://en.wikipedia.org/w/index.php?title=File:I-76_(CO).svg *License*: unknown *Contributors*: Ltljltlj, T2, 2 anonymous edits

File:US 87.svg *Source*: http://en.wikipedia.org/w/index.php?title=File:US_87.svg *License*: unknown *Contributors*: Bidgee, SPUI, Xnatedawgx, 3 anonymous edits

File:US 400.svg *Source*: http://en.wikipedia.org/w/index.php?title=File:US_400.svg *License*: unknown *Contributors*: SPUI, Xnatedawgx

File:I-225 (CO).svg *Source*: http://en.wikipedia.org/w/index.php?title=File:I-225_(CO).svg *License*: unknown *Contributors*: User:Ltljltlj

File:US 36.svg *Source*: http://en.wikipedia.org/w/index.php?title=File:US_36.svg *License*: unknown *Contributors*: Bidgee, SPUI, Xnatedawgx, 3 anonymous edits

File:US 491.svg *Source*: http://en.wikipedia.org/w/index.php?title=File:US_491.svg *License*: unknown *Contributors*: SPUI

File:I-270 (CO).svg *Source*: http://en.wikipedia.org/w/index.php?title=File:I-270_(CO).svg *License*: unknown *Contributors*: Ltljltlj, 2 anonymous edits

File:US 40.svg *Source*: http://en.wikipedia.org/w/index.php?title=File:US_40.svg *License*: unknown *Contributors*: Rocket000, SPUI

File:US 160.svg *Source*: http://en.wikipedia.org/w/index.php?title=File:US_160.svg *License*: unknown *Contributors*: AFBorchert, SPUI

File:US 550.svg *Source*: http://en.wikipedia.org/w/index.php?title=File:US_550.svg *License*: unknown *Contributors*: AFBorchert, SPUI, Xnatedawgx

File:US 50.svg *Source*: http://en.wikipedia.org/w/index.php?title=File:US_50.svg *License*: unknown *Contributors*: Rocket000, SPUI

File:US 285.svg *Source*: http://en.wikipedia.org/w/index.php?title=File:US_285.svg *License*: unknown *Contributors*: SPUI, 1 anonymous edits

File:Denver International Airport security.jpg *Source*: http://en.wikipedia.org/w/index.php?title=File:Denver_International_Airport_security.jpg *License*: unknown *Contributors*: User:Dbenbenn

File:UnionStationDenver01 gobeirne.jpg *Source*: http://en.wikipedia.org/w/index.php?title=File:UnionStationDenver01_gobeirne.jpg *License*: unknown *Contributors*: User:gobeirne

File:Denver capital.jpg *Source*: http://en.wikipedia.org/w/index.php?title=File:Denver_capital.jpg *License*: unknown *Contributors*: User:Hustvedt

image:Map of Colorado counties, labelled.svg *Source*: http://en.wikipedia.org/w/index.php?title=File:Map_of_Colorado_counties,_labelled.svg *License*: unknown *Contributors*: User:Dbenbenn

File:Colorado census statistical areas.svg *Source*: http://en.wikipedia.org/w/index.php?title=File:Colorado_census_statistical_areas.svg *License*: unknown *Contributors*: Buaidh

File:2006-07-14-Denver Skyline Midnight.jpg *Source*: http://en.wikipedia.org/w/index.php?title=File:2006-07-14-Denver_Skyline_Midnight.jpg *License*: unknown *Contributors*: Matt Wright

File:Magnify-clip.png *Source*: http://en.wikipedia.org/w/index.php?title=File:Magnify-clip.png *License*: unknown *Contributors*: User:Erasoft24

image:Cadet chapel.jpg *Source*: http://en.wikipedia.org/w/index.php?title=File:Cadet_chapel.jpg *License*: unknown *Contributors*: Original uploader was Tarfu92 at en.wikipedia

image:Norlin Library - Colorado.jpg *Source*: http://en.wikipedia.org/w/index.php?title=File:Norlin_Library_-_Colorado.jpg *License*: unknown *Contributors*: User:MECU

image:University of Denver campus pics 003.jpg *Source*: http://en.wikipedia.org/w/index.php?title=File:University_of_Denver_campus_pics_003.jpg *License*: unknown *Contributors*: Original uploader was CW221 at en.wikipedia

image:PetersonAFB.jpg *Source*: http://en.wikipedia.org/w/index.php?title=File:PetersonAFB.jpg *License*: unknown *Contributors*: Common Good, Sascha Brück, Xnatedawgx, 1 anonymous edits

image:Coloradodunes.jpg *Source*: http://en.wikipedia.org/w/index.php?title=File:Coloradodunes.jpg *License*: unknown *Contributors*: Dbenbenn, Ebyabe, Huebi, Man vyi, Rmhermen, TomCatX, Urban, Uroboros

image:SpruceTreeHouseLong.jpg *Source*: http://en.wikipedia.org/w/index.php?title=File:SpruceTreeHouseLong.jpg *License*: unknown *Contributors*: Scott Catron

image:Coors field 1.JPG *Source*: http://en.wikipedia.org/w/index.php?title=File:Coors_field_1.JPG *License*: unknown *Contributors*: Lipothymia, Moofpocket, Silver Spoon, Spyder Monkey, UCinternational, 2 anonymous edits

image:Denver invesco stadium 1.jpg *Source*: http://en.wikipedia.org/w/index.php?title=File:Denver_invesco_stadium_1.jpg *License*: unknown *Contributors*: KM Newnham

image:Pepsi Center.jpg *Source*: http://en.wikipedia.org/w/index.php?title=File:Pepsi_Center.jpg *License*: unknown *Contributors*: Sascha Brück, Spyder Monkey, Verne Equinox, Xgeorg

File:Harlan_County_Nebraska_Incorporated_and_Unincorporated_areas_Republican_City_Highlighted.svg *Source*: http://en.wikipedia.org/w/index.php?title=File:Harlan_County_Nebraska_Incorporated_and_Unincorporated_areas_Republican_City_Highlighted.svg *License*: unknown *Contributors*: Arkyan

File:Harlan_County_Nebraska_Incorporated_and_Unincorporated_areas_Alma_Highlighted.svg *Source*: http://en.wikipedia.org/w/index.php?title=File:Harlan_County_Nebraska_Incorporated_and_Unincorporated_areas_Alma_Highlighted.svg *License*: unknown *Contributors*: Arkyan

File:Carson house (Brownville, Nebraska) from NE.JPG *Source*: http://en.wikipedia.org/w/index.php?title=File:Carson_house_(Brownville,_Nebraska)_from_NE.JPG *License*: unknown *Contributors*: User:Ammodramus

File:Nemaha_County_Nebraska_Incorporated_and_Unincorporated_areas_Brownville_Highlighted.svg *Source*: http://en.wikipedia.org/w/index.php?title=File:Nemaha_County_Nebraska_Incorporated_and_Unincorporated_areas_Brownville_Highlighted.svg *License*: unknown *Contributors*: Arkyan

MIX
Papier aus verantwortungsvollen Quellen
Paper from responsible sources
FSC® C105338

Printed by Books on Demand GmbH, Norderstedt / Germany